Networked and Event-Triggered Control Approaches in Cyber-Physical Systems

Networked and Event-Triggered Control Approaches in Cyber-Physical Systems

Jinhui Zhang
Yuanqing Xia
Zhongqi Sun
Duanduan Chen

CRC Press
Taylor & Francis Group
Boca Raton London New York

CRC Press is an imprint of the
Taylor & Francis Group, an **informa** business

First edition published 2022
by CRC Press
6000 Broken Sound Parkway NW, Suite 300, Boca Raton, FL 33487-2742

and by CRC Press
2 Park Square, Milton Park, Abingdon, Oxon, OX14 4RN

CRC Press is an imprint of Taylor & Francis Group, LLC

ISBN: 978-1-032-19794-4 (hbk)
ISBN: 978-1-032-19795-1 (pbk)
ISBN: 978-1-003-26088-2 (ebk)

DOI: 10.1201/9781003260882

Publisher's note: This book has been prepared from camera-ready copy provided by the authors.

Dedicated to our parents

Contents

List of Figures   xi

List of Tables   xv

Preface   xvii

CHAPTER 1 ▪ Introduction   1

1.1	CONCEPTS AND CHALLENGES OF CPS	1
1.2	NETWORKED CONTROL APPROACHES IN CPS	2
1.3	EVENT-TRIGGERED CONTROL STRATEGIES IN CPS	5
1.4	OUTLINE OF THIS BOOK	8

SECTION I   Networked Control Approaches in CPSs

CHAPTER 2 ▪ Stochastic Stability and Stabilization of CPSs with Random Delay   13

2.1	INTRODUCTION	13
2.2	SYSTEM DESCRIPTION	14
2.3	MODELING OF CPS	15
2.4	STOCHASTIC STABILITY AND STABILIZATION	16
2.5	NUMERICAL SIMULATION	20
2.6	SUMMARY	23

CHAPTER 3 ▪ Observer-Based Output-Feedback Stabilization for CPSs with Quantized Inputs   25

3.1	INTRODUCTION	25
3.2	SYSTEM DESCRIPTION	26
3.3	FULL-ORDER OBSERVER-BASED QUANTIZED CONTROL	28
3.4	REDUCED-ORDER OBSERVER-BASED QUANTIZED CONTROL	31

3.5 NUMERICAL SIMULATION 33

3.6 SUMMARY 35

CHAPTER 4 ■ State-Feedback Networked Delay Compensation Control for CPSs with Time-Varying Delay 37

4.1 INTRODUCTION 37

4.2 SYSTEM DESCRIPTION 38

4.3 DELAY COMPENSATION CONTROL SCHEME 39

4.4 STABILITY ANALYSIS 41

4.5 PRACTICAL EXPERIMENT 46

4.6 SUMMARY 50

CHAPTER 5 ■ Output-Feedback Networked Delay Compensation Control for CPSs with Random Delay 51

5.1 INTRODUCTION 51

5.2 SYSTEM DESCRIPTION 52

5.3 STOCHASTIC DELAY COMPENSATION CONTROL SCHEME 53

5.4 STOCHASTIC STABILITY ANALYSIS 54

5.5 CONTROLLER SYNTHESIS 57

5.6 NUMERICAL SIMULATION 62

5.7 SUMMARY 65

CHAPTER 6 ■ Fuzzy Networked Delay Compensation Control for Nonlinear CPSs 67

6.1 INTRODUCTION 67

6.2 SYSTEM DESCRIPTION 68

6.3 FUZZY DELAY COMPENSATION CONTROL SCHEME 69

6.4 STABILITY ANALYSIS AND CONTROLLER DESIGN 71

6.5 NUMERICAL SIMULATION 77

6.6 SUMMARY 80

CHAPTER 7 ■ Networked Output Tracking Control of CPSs with Delay Compensation 81

7.1 INTRODUCTION 81

7.2   SYSTEM DESCRIPTION   82

7.3   DELAY COMPENSATION IN SENSOR-TO-CONTROLLER
      NETWORK CHANNEL   83

7.4   DELAY COMPENSATION IN CONTROLLER-TO-ACTUATOR
      NETWORK CHANNEL   87

7.5   STABILITY AND STEADY-STATE TRACKING ERROR ANALYSIS   89

7.6   NUMERICAL SIMULATION   92

7.7   SUMMARY   93

SECTION II Event-Triggered Control Approaches in CPSs

CHAPTER 8 ▪ Observer-Based Event-Triggered Control
            for CPSs   97

8.1   INTRODUCTION   97

8.2   SYSTEM DESCRIPTION   98

8.3   EVENT-TRIGGERED CONTROL WITH CONTINUOUS UIS   99

8.4   EVENT-TRIGGERED CONTROL WITH DISCRETE-TIME EVENT DE-
      TECTOR   103

8.5   NUMERICAL SIMULATION   106

8.6   SUMMARY   109

CHAPTER 9 ▪ Observer-Based Self-Triggered Control for CPSs   111

9.1   INTRODUCTION   111

9.2   SYSTEM DESCRIPTION   112

9.3   FULL-ORDER OBSERVER-BASED SELF-TRIGGERED CONTROL   113

9.4   REDUCED-ORDER OBSERVER-BASED SELF-TRIGGERED
      CONTROL   117

9.5   NUMERICAL SIMULATION   121

9.6   SUMMARY   124

CHAPTER 10 ▪Event-Triggered Dynamic Output-Feedback Control for CPSs 125

10.1   INTRODUCTION   125

10.2   SYSTEM DESCRIPTION   126

10.3   EVENT-TRIGGERED DOF CONTROL FOR SYSTEMS WITH
       CONTINUOUS OUTPUT MEASUREMENTS   126

10.4 EVENT-TRIGGERED DOF CONTROL FOR SYSTEMS WITH SAMPLED OUTPUT MEASUREMENTS   130

10.5 NUMERICAL SIMULATION   134

10.6 SUMMARY   137

CHAPTER 11 ▪Co-Design of Event-Triggered Control and Quantized Control for CPSs   139

11.1 INTRODUCTION   139

11.2 SYSTEM DESCRIPTION   140

11.3 QUANTIZED EVENT-TRIGGERED CONTROL: FIXED EVENT THRESHOLD STRATEGY   141

11.4 QUANTIZED EVENT-TRIGGERED CONTROL: RELATIVE THRESHOLD STRATEGY   147

11.5 NUMERICAL SIMULATION   150

11.6 SUMMARY   152

CHAPTER 12 ▪Event-Triggered Disturbance Rejection Control for CPSs   153

12.1 INTRODUCTION   153

12.2 SYSTEM DESCRIPTION   154

12.3 FULL-ORDER ESO-BASED EVENT-TRIGGERED DISTURBANCE REJECTION CONTROL   155

12.4 REDUCED-ORDER ESO-BASED EVENT-TRIGGERED DISTURBANCE REJECTION CONTROL   161

12.5 REDUCED-ORDER EFSO-BASED EVENT-TRIGGERED DISTURBANCE REJECTION CONTROL   166

12.6 NUMERICAL SIMULATION AND EXPERIMENTS   171

12.7 SUMMARY   180

Bibliography   181

Index   199

List of Figures

2.1   State and control input signals of aircraft system (state-feedback case)   21

2.2   State and control input signals of aircraft system (output-feedback case)   22

3.1   Logarithmic quantizer   26

3.2   Block diagram of observer-based quantized feedback control   27

3.3   Satellite system   33

3.4   State and control input signals of satellite system (full-order case)   34

3.5   State and control input signals of satellite system (reduced-order case)   35

4.1   Block diagram of state-feedback networked delay compensation control   38

4.2   Ball and beam system and its schematic diagram   47

4.3   Networked control test rig   48

4.4   Comparative experiment results for the position of the ball (case 1)   49

4.5   Comparative experiment results for the position of the ball (case 2)   50

5.1   Block diagram of output-feedback networked delay compensation control   52

5.2   States and control input of motor control system (known transition probabilities case)   63

5.3   States and control input of motor control system (partially known transition probabilities case)   64

5.4   States and control input of motor control system (unknown transition probabilities case)   65

6.1   Block diagram of fuzzy networked delay compensation control   68

6.2   Inverted pendulum controlled by a DC motor   77

6.3   States and control input of inverted pendulum system (state-feedback case)   78

6.4   States and control input of inverted pendulum system (output-feedback case)   79

7.1   Block diagram of networked tracking control systems   82

7.2   Comparative outputs of the motor control system and network delays
       $(r(k) = r_1(k))$ 93

7.3   Comparative outputs of the motor control system and network delays
       $(r(k) = r_2(k))$ 94

8.1   Block diagram of event-triggered control system   98

8.2   Inverted pendulum system   106

8.3   Comparative simulation results of time- and event-triggered control
       approaches (continuous UIS case)   107

8.4   Comparative simulation results of time- and event-triggered control
       approaches (discrete UIS case)   108

9.1   Block diagram of self-triggered control system   112

9.2   Comparative simulation results of time- and self-triggered control ap-
       proaches (full-order observer case)   122

9.3   Comparative simulation results of time- and self-triggered control ap-
       proaches (reduced-order observer case)   123

10.1   Illustration of discrete UIS   131

10.2   Comparative simulation results of time- and event-triggered DOF
       control approaches (the continuous output case)   135

10.3   Comparative simulation results of time- and event-triggered DOF
       control approaches (the sampled output case)   136

11.1   Block diagram of quantized event-triggered control systems   140

11.2   Structure of the event-triggered encoder   142

11.3   Structure of the decoder   143

11.4   Schematic model of PVTOL aircraft   150

11.5   Simulation results of quantized event-triggered control approach
       (fixed event threshold case)   151

11.6   Simulation results of quantized event-triggered control approach (rel-
       ative event threshold case)   152

12.1   Comparative simulation results of event-triggered disturbance rejec-
       tion control approach (full-order observer case)   172

12.2   Comparative simulation results of event-triggered disturbance rejec-
       tion control approach (reduced-order ESO case)   174

12.3   Simulation results of the reduced-order EFSO-based event-triggered
       control approach (matched disturbance case)   176

12.4   Simulation results of the system with mismatched disturbance $f(k)$ 177

12.5 The rotary inverted pendulum and its schematic diagram   178

12.6 Experiment results for the reduced-order EFSO-based event-triggered
control approach   179

List of Tables

2.1   Probability distribution of random network delays and data dropouts   21

2.2   Allowable upper bounds of d_M for various d_m 23

2.3   Comparison with delay-partitioning approach   23

4.1   The parameters of beam and ball system   47

12.1   Updating times of controller for different Δ_i (matched disturbance case)   175

12.2   Updating times of controller for different Δ_i (mismatched disturbance case)   177

List of Tables

Preface

With the development of the science and technology, modern control systems are becoming more and more complex. Due to the limitation of space area, traditional local control approaches cannot satisfy the actual requirements, gradually. At the right moment, the networked embedded system is born to deal with the spatial control problems. Then, on the basis of the networked embedded system, cyber-physical systems (CPSs) are proposed as the extension in both size and complexity through control, computation, and communication together. The main advantages of CPSs are low cost, reduced weight and power requirements, simple installation and maintenance, as well as high reliability. However, it should be pointed out that with the utilization of communication networks, some challenging issues, such as network communication delay, data dropout, signal quantization, event-triggered communication, and so on, should be taken into account. In view of these challenges, the networked and event-triggered control approaches are researched in this book.

Due to the limited bit rate of the communication channels, node waiting time to send out a packet via a busy network channel, or because the time for signal processing and propagation is high, the network communication delay is unavoidable. In CPSs, the delays may occur in both the sensor-to-controller and controller-to-actuator network channels. Besides, the network congestion, bit transmission error, link failure, and so on can easily lead to data dropouts. In order to avoid the network congestion and reduce the bandwidth of the network, one effective way is to quantize the signals before transmission, and then the real-valued signals are quantized as the piecewise signals; however, the ideal control performance will be affected by the quantization errors. Moreover, the CPSs subject to network communication delay or data dropouts can also be modeled as the time-delay, switched, or stochastic systems, then the stability conditions of the resulting closed-loop systems can be developed by applying the time-delay, switched, or stochastic system analysis approaches. It is worth mentioning that the time-delay, switched, or stochastic system approaches simply passively accept the presence of the network communication delay and data dropout. To actively compensate the performance losses caused by network communication delay or data dropouts, the networked delay compensation control approach is developed by taking full advantage of the packet-based transmission in CPSs. In another way, to reduce the utilization rate of communication resources, the event-triggered instead of time-triggered communication approach is adopted, which can maintain the control performance while improving the communication quality. Typical event-triggered control approaches include continuous event-triggered control, periodic event-triggered control, self-triggered control, and dynamic event-triggered control strategies. For the continuous event-triggered control strategy, it is essential to discuss the avoidance of Zeno behavior, which is a response of the system when an infinite number of triggerings happen in a finite amount of time. To avoid the continuous monitoring of the event-triggering condition, the periodic event-triggered control strategy is proposed, and in the self-triggered control strategy, only the state measurements at the current updating instant are utilized to determine the next updating instant. Furthermore, the dynamic event-triggered control approach is proposed by introducing an additional internal dynamical variable to enlarge the minimum inter-event interval and further reduce the usage of the communication resources.

During the past 30 years, networked control problems have already caught the attentions of many international scholars owing to the great importance of CPSs. In this book, the authors aim to introduce networked and event-triggered control approaches to cope with the compensation control problems for systems with network communication delay, data dropouts, signal quantization, and event-triggered communications. The contributions of this book include both theoretic innovations and practical applications. In the theoretic aspect, this book mainly proposes two appealing networked control approaches including the networked delay compensation control and event-triggered control. The state-/output-feedback networked delay compensation control approaches are proposed for linear/nonlinear physical plants by applying the time-delay system, switched system, or stochastic system analysis approaches. To reduce the usage of network resources, the quantized, event-/self-triggered control approaches are presented for systems with signal quantization/unknown external disturbances. In the practical aspect, the application examples, such as satellite systems, the ball and beam system, motor control systems, the inverted pendulum system, and so on, are utilized to verify the theoretic results of each chapter through either simulations or experiments. Furthermore, it is interesting to extend the researched networked and event-triggered control approaches into more complex environments, such as unmanned vehicle systems, cloud control systems, and so on.

The major topics of this book are about theoretic methods and practical applications of networked and event-triggered control approaches in CPSs. The book is arranged as the following sections: Chapter 1 introduces the background and challenges of CPSs, the basic control approaches, and the motivations. Section I, which is composed of Chapters 2–7, provides the networked control approaches for CPSs. In Chapter 2, by considering the random characteristics of network communication delay and data dropout, the modeling approach is developed based on the probability distribution of the network communication delay and data dropout. In Chapter 3, the observer-based output-feedback quantized control problem is studied for discrete-time linear systems by using a polytopic approach. Chapter 4 addresses the problem of networked delay compensation control and stability analysis for systems with time-varying network communication delay. In Chapter 5, the design problem of a static output-feedback networked delay compensation controller is considered for systems with random network communication delay. Chapter 6 is concerned with the network delay compensation problem for a nonlinear plant represented by the Takagi-Sugeno (T-S) fuzzy model. In Chapter 7, the problem of networked output tracking control is investigated by considering the delay compensations in both the sensor-to-controller and controller-to-actuator channels in CPSs. Moreover, Chapters 8–12, which are integrated in Section II, discuss the event-triggered control problem in CPSs. Chapter 8 concerns the problem of observer-based event-triggered output-feedback control of continuous-time systems. In Chapter 9, the authors study the self-triggered output-feedback control problem of discrete-time systems. Chapter 10 focuses on designing dynamic output-feedback controllers for systems with event-triggered control inputs. Then, the co-design problem of event-triggered control and quantized control techniques is solved in Chapter 11. To cope with the external disturbance, the event-triggered disturbance compensation control approaches are studied in Chapter 12.

To adequately read and follow the content of this book, only the knowledge of basic automatic control theory is required for readers. We believe that this book will be of interest to university researchers, engineers, and graduate and undergraduate students in mechanical, control, computer, and communication engineering, and the people who are interested in learning the networked, event-triggered control approaches and their applications in CPSs.

ACKNOWLEDGMENTS

The main contents of this book are the result of extensive research efforts in the past ten years. Some of the results have been reported in the research papers. We would like to express our thanks to our collaborators, Peng Shi, Wei Xing Zheng, James Lam, and Gang Feng, for their active discussions and valuable suggestions when we carry out research in this area. Our thanks also go to the former and current students, Danyang Zhao, Yujuan Lin, and Xinwei Liu from the College of Information Science & Technology at Beijing University of Chemical Technology, and Hao Xu, Kai Zhao, Shaomeng Gu, Xin Liu, Sihang Li, and Huan Meng from the School of Automation at Beijing Institute of Technology, for their diligent and active works on the achievements of this book.

Most parts of this book are supported in part by the National Key Research & Development Program of China under Grant 2018YFB1700602, in part by the National Natural Science Foundation of China under Grant 61873030, Grant 61973230, Grant 61473024, and Grant 61104098.

Beijing

Jinhui Zhang
Yuanqing Xia
Zhongqi Sun
Duanduan Chen

CHAPTER 1

Introduction

1.1 CONCEPTS AND CHALLENGES OF CPS

In the wake of the rapid developments in computer and communication technologies, network-based communication and control technologies have played important roles in many industrial control applications, such as manufacturing plants, power plants, automobiles, aircrafts, robot manipulators, and so on. The cyber-physical systems (CPSs) are new class of networked embedded systems, and tightly integrate computation, networking, and physical processes. Different from the traditional point-to-point system architecture, CPSs are spatially distributed systems, and the signals between the system components are transmitted via a communication network, such as the remote control of several mobile units, arrays of micro-actuators, and underwater acoustics.

In order to cope with the uncertainties, complexities and resource limitations of network communications in practical applications, the co-design between control and communication should be considered. Networked control and event-triggered control are the two popular control approaches in CPSs. Although fruitful results on networked control and event-triggered control can be found in recent publications, there are still a lot of spaces for further investigation by considering complexity of CPSs in applications, and it is essential to continue in-depth researches on the CPSs.

In spite of the benefits, the insertion of communication networks in feedback control loops makes the analysis and synthesis of CPSs complex, and the characteristics of communication network, such as network communication delay, data dropout, signal quantization, may degrade the system control performances. In CPSs, these challenging network-induced issues should be taken into account owing to the communication constraints.

- **Network communication delay.** The network communication delay is usually caused by limited bit rate of the communication channels, node waiting to send out a packet via a busy channel, or signal processing and propagation. Generally, two kinds of network communication delays in CPSs should be considered, one is the delay occurs in the sensor-to-controller network channel and the other is the delay occurs in the controller-to-actuator network channel. It should be pointed out that these two kinds of delays may possess different

DOI: 10.1201/9781003260882-1

characteristics [114]; however, in [71, 42, 121, 187, 207], to perform the stability analysis and controller design, these delays are treated as round trip delay for simplicity.

- **Data dropout.** Data dropout phenomenon is usually caused by the network congestion, bit transmission error, link failure, and so on. On the one hand, the data dropout occurs if the transmitted data packets cannot be successfully received by their destination. On the other hand, the data packets will be discarded if they cannot arrive at a destination within a certain transmission time according to the commonly used network protocols, and the discarded data packet will also be classified as data dropout. The data dropout will significantly degrade the control performance of CPSs [163, 176, 217, 206, 14, 16].

- **Signal quantization.** Owing to the limited bandwidth of the network, signals should be quantized before transmission. The key idea of signal quantization in CPSs is to convert a real-valued signal into a piecewise constant one taking on a finite set of values. Additionally, the signal quantization is also a potential method to reduce the bandwidth and avoid the network congestion. The quantization error, the difference between the original signal and the quantized signal, also have significant impacts on the system performance [25, 27, 97, 35, 40, 199].

- **Event-triggered communication.** In the CPSs, the time-triggered and event-triggered communication strategies are usually adopted. The time-triggered communication implies that all the periodically sampled signals are transmitted over network, which may lead to unnecessary utilization of the limited network resources. In the event-triggered communication scenario, to guarantee desired specifications of the systems, only the necessary signals are transmitted over network, and the signal transmissions are decided by the occurrences of a predefined "event." The event-triggered communication schemes can maintain the control performance while saving the limited communication resources [109, 3, 67, 118, 189, 198].

1.2 NETWORKED CONTROL APPROACHES IN CPS

In recent decades, the networked control approaches and analysis of CPSs have received substantial attention from research communities, and research topics mainly include modeling, stability analysis, control, and filtering problems. See [42, 41, 44, 181, 182, 207, 13, 232, 14, 167, 201, 196, 63, 18, 81, 177, 176, 163, 200, 186, 191, 190, 187, 152, 166, 120, 146, 204, 79, 75, 217] and the references therein. For more details, please refer to the special issue [5], survey paper [56], and monographes [153, 78, 171, 124, 184, 224, 117]. Meanwhile, many interesting analysis and control approaches have been proposed to cope with the network-induced problems, among which the following main approaches are introduced.

- **Time-delay system approach.** The time-delay system approach is developed from the input delay approach for sampled-data systems [34, 186], and the networked closed-loop system is modeled as a system with a time-varying input delay. A number of results are established by applying the Lyapunov-Krasovskii function method, or its modifications by adding some terms including information on the lower bound and the upper bound of time-varying delay [34, 186]. In [66], the delay-dependent stability condition is established by using a method of estimating the upper bound on the difference of a Lyapunov function without ignoring any terms. In [110], the delay partition method is proposed for establishing the delay-dependent stability condition for linear discrete-time systems. Also by applying the delay partition approach, a delay-dependent sufficient condition is proposed for Takagi-Sugeno (T-S) fuzzy time-delay systems in [168, 169]. In the aforementioned results, the information on the sensor-controller delay, the controller-actuator delay, and the number of consecutive packet dropouts is implicitly included in the upper bound of the input delay. However, the maximum delay upper bound indicates the worst case of network communication delays and the number of consecutive packet dropouts that the system can tolerate. If the worst case occurs rarely, the results derived from the time-delay system approach are certainly conservative, which is shown through some examples in [44]. The work along this direction is to derive a less conservative result by utilizing much more delay information [38, 42]. A time-varying Lyapunov functional proposed in [33] is shown to be an effective one to reduce the design conservatism as the functional is time-dependent. In [121], an improved interval-delay-dependent Lyapunov functional is proposed and an extended Jessen's inequality is introduced, which shows that the maximum allowable delay bound and the feedback gain of a memoryless controller could be derived by solving a set of linear matrix inequalities (LMI).

- **Switched system approach.** Another important tool for modeling, analysis, and design of networked plant is switched system, which is typically used by modeling different network conditions in CPSs as different system modes [212, 214]. Under the assumption that the control signal is time-varying within a sampling period, a discrete-time switched model is established in [98] by dividing the sampling period into a number of subintervals corresponding to the controller's reading buffer period, and within each subinterval, the control signal is assumed to be constant. By employing the average dwell time technique, some quantitative relations between the packet dropout rate and stability are established [213]. In order to remove the assumption that the network communication delay is less than one sampling period, a lot of efforts are made in [214, 24, 72]. In [214], the system with network communication delays is modeled as a discrete-time switched delay system with both stable subsystems and unstable subsystems. By using an event-based discrete-time representation of the continuous-time physical plant, the networked system is modeled as a switched system with arbitrary switching [72], in which a long network communication delay and nonuniform sampling are allowed. Practically, the physical plant is equipped with sensors and actuators that are grouped into a number of nodes [24], and at each transmission instant, a switching indicator is used to choose only one certain node to access the network and transmit its corresponding data packets, and then a discrete-time switched linear uncertain system model is established. In the continuous-time domain, a switched system approach is also studied for system with network communication delays in [87] and [142]. By introducing a switching function related to the variation in network communication delays, the closed-loop system is modeled as a switched time-delay system with two switching modes, and each mode has a different controller gain. Together with the average dwell time technique, both stability and exponential stabilization are investigated in [142] and [87], respectively.

- **Stochastic system approach.** Generally, in CPSs, the random demands of the utilization of a multi-user network may affect the network traffic, and result in random delays and/or dropout of transmitted data, and the system performance as well as stability will be deteriorated. Thus, the stochastic system approaches are developed for modeling and analysis of the networked systems with random delay and/or data dropout. A networked strong tracking filter is proposed in [65] for a class of nonlinear networked systems with multiple packet dropout modeled by Bernoulli stochastic process, and desired estimation performance is achieved. In [64], a unified measurement model is utilized to simultaneously characterize both the phenomena of multiple communication delays and data dropout, based on which the fault detection and isolation problem is effectively solved by using a direct state estimation approach. According to [177], the controller always uses the most recent data, and the data dropout can be treated as network delay case, and the network delay can increase at most by one unit at each step. By modeling the random delay as Markov chains, the augmented closed-loop system is established as a Markov jump linear system, and the stabilization problem is considered in [174], and this idea is further extended to use two Markov chains to model the delays in both feedback and forward channels in [209, 137]. Moreover, in some situations, such as those systems connected over a wireless network, the observation of communication delay of the transmitted data includes the probability distribution of the delay values, which implies the case that the possible value of the delay with a low probability can be very large. By making full use of the information concerning the probability distribution of the delays, the stochastic stability and stabilization controller design are investigated for networked system with probabilistic interval input delays in [190].

- **Networked delay compensation approach.** It should be noted that, in the aforementioned works, the system simply passively accepts the presence of the network communication delay and data dropout. One efficient solution for compensating network communication delay is the so-called "networked predictive control (NPC)" or "packet-based control," which is first proposed by Liu et al. [99] and then improved by Zhao [228, 223, 226, 225, 227]. The key idea is sending a sequence of control input predictions in one data packet, and if a new data packet arrives at actuator side, a comparison process in the network delay compensator is introduced, since the newly arrived data packet may contain the old data because of the presence of time-varying communication delays.

The network delay compensator will compare the time stamps of the arrived data packet with the one stored in it, and will be updated only when the arrived data packet is more recent than the existing one. That is, the network delay compensator only accept the arrived data packet if its time stamp value is greater than that of the packet stored in the network delay compensator currently. As a result of the comparison process, the data packet stored in the network delay compensator is always the latest one available at any specific time. After the comparison process, the suitable control input in the data packet stored in network delay compensator is selected to actively compensate the network communication delay. Further studies on NPC methods are made in [100, 101, 228, 154, 156, 207, 204, 55]. For example, in [225, 226], some NPC approaches are proposed for networked Hammerstein-type systems. In [100, 101], the state space model based NPC methods is developed for networked control systems (NCSs) with feedback channel delay and both forward and feedback channel delays, respectively. In [126], the NPC approach is developed for uncertain constrained nonlinear systems. In [77], an event-driven NPC method is presented for single input single output (SISO) networked systems.

1.3 EVENT-TRIGGERED CONTROL STRATEGIES IN CPS

It is well-known that in conventional sampled-data systems, the sampling of the outputs, the computing of the control laws and the updating of the actuator are carried out with a constant time period, which is called time-triggered scheme, and has been widely investigated in the past decades [15, 16, 32, 33, 46, 73, 121, 135, 180, 216, 231]. In CPSs, communication networks are employed to exchange information among the sensors, controllers, and actuators, and the time-triggered control/communication scheme can increase the network communication loads which may lead to network congestion, time delay or packet dropouts, and thus degrades the system performance or even destabilizes the system at certain conditions. On the premise that the desired control performance can be guaranteed, it is necessary to decrease possible network congestion by reducing the communication loads. Event-triggered (also called deadband scheduling) approaches [6, 8, 70, 74, 83, 106, 116, 131, 157, 189, 198, 220] have been witnessed as an effective manner to reduce the loads of communication networks. The main idea behind the event-triggered control/communication scheme is that the necessary transmissions of system and executing of control inputs are decided by the occurrence of a predetermined "event" while ensuring the desired control performances. Compared with the time-triggered control/communication scheme, the event-triggered control/communication strategy not only decreases the computational burden of the controller but also reduces the loads of the network communication. Mathematically speaking, in an event-triggered control system, the event detector is always implemented to monitor if the predefined event triggering condition is violated or not. Generally, the existing event-triggered control strategies can be roughly classified into four categories, including continuous event-triggered control strategy, periodic event-triggered control strategy, self-triggered control strategy, and dynamic event-triggered control strategy. One of the important issues in designing event-triggered control systems, is the so called Zeno behavior, which is a response of system when infinite number of triggering happen in a finite amount of time. The Zeno behavior is clearly undesirable from an implementation point of view. Therefore, it is crucial in an event-triggered control system that the designed event triggering condition guarantees a lower bound for the time intervals between triggering instants, and consequently ensures that there is no Zeno behavior in the system. This lower bound is also referred as minimum inter-event time.

- **Continuous event-triggered control strategy.** In this case, the event detector monitors the event-triggering condition continuously. In [106], event-triggered state-feedback control is considered for linear systems with bounded disturbances, the results are extended to event-triggered control with communication delays and packet losses in [91], and more results can be found in the monograph [88]. The minimum inter-event time for event-triggered control systems is determined in [143] by enforcing input-to-state stability of the system. Following a similar line of [143], the event-triggered policy is resorted for wireless sensor/actuator networks in [109], which results in less energy consumption. Moreover, it is worth mentioning that the full state information is not available in many control applications, so it is important to investigate the output feedback based event-triggered control strategies. In [185], the output feedback based event-triggered control framework is proposed for CPSs, and the event triggering condition is derived based on the passivity theory. In [90], the event-triggered state-feedback control approach established in [106] is extended toward event-based output feedback control, where a state observer is incorporated to generate the control inputs. In [23], the stability and $\mathcal{L}_\infty$ performance of event-triggered control systems with dynamical output feedback controllers are studied with decentralized event-triggered mechanisms.

- **Periodic event-triggered control strategy.** Although the continuous event-triggered control strategy provides a useful way to reduce the loads of the network communication, it usually requires continuous monitoring of the event-triggering condition. For avoiding the continuous monitoring, the periodic event-triggered control strategies have been proposed in [6, 67, 68, 69, 118, 123, 189, 220], where sensors only sample the plant states with a constant rate and transfer the sampled information to event generator which determines whether the event should be triggered or not. Therefore, the event-triggered conditions in the periodic event-triggered control strategy are no longer needed to be monitored continuously. In [67, 69], a general framework for periodic event-triggered control is proposed, and three different approaches including impulsive systems, piecewise linear systems and perturbed linear systems, are presented to analyze the stability and L_2-gain properties of the periodic event-triggered control systems. In [189], a delay system approach is proposed to model the continuous system with event-triggered controller where the event-triggering condition is established based on the periodically sampled states. The similar approaches are proposed to deal with sampled-data control systems in [118] and H_∞ control co-design for CPSs in [123], respectively.

- **Self-triggered control strategy.** The event-triggered conditions in the aforementioned two strategies rely on continuous or periodic supervision of the system states in order to detect whether the event triggering condition is violated or not, which may result in the complexity of implementation of the event-triggered control strategies. To tackle this issue, the self-triggered control strategy is proposed in [2, 3], where only the state measurements at the current sampling instant need to be known for determining the next sampling instant. In [157], a self-triggered scheme is presented for linear time-invariant systems with the disturbance, where the magnitude of the disturbance is bounded by a linear function of the norm of the system state, and the finite-gain L_∞ stability of the resulting self-triggered feedback control systems is guaranteed. In [158], the L_∞ stability of self-triggered feedback systems with state-independent disturbances is further investigated. Besides, the self-triggered state and output feedback control strategies are proposed for system with disturbances in [2] and [3], respectively. In [4], a self-triggered control scheme is proposed for both nonlinear state-dependent homogeneous systems and polynomial systems.

- **Dynamic event-triggered control strategy.** Different from the aforementioned three static event-triggered control strategies, a dynamic event-triggered control strategy is proposed in [51] in order to further reduce the communication and energy resources. It should be noted that for the dynamic event-triggered control strategy, by introducing an additional internal dynamical variable, the minimum inter-event time is enlarged and the number of triggering is reduced in general [161, 48, 159, 20, 94]. In view of this appealing advantage, a lot of works have been done to deal with the communication resource-saving problem in CPSs by now [93, 7, 36, 192]. In [150], for the linear uncertain system, a robust controller is designed with a dynamic event-triggering mechanism to reduce the communication load. In the presence of sporadic measurements and communication delays, two dynamic discrete-time event-triggered control strategies are utilized for the linear stochastic systems in [107]. For the nonlinear stochastic systems, both the dynamic event-triggered and self-triggered control approaches are proposed and the stochastic stability of the closed-loop system is ensured [160]. In [76], an adaptive event-triggering scheme is presented for discrete-time nonlinear systems. Considering the nonlinear uncertain systems subject to additive time-varying disturbance, an output-based dynamic event-triggered mechanism is provided for disturbance rejection control in [141]. In [183], two distributed dynamic triggering laws are proposed to solve the consensus problem for multi-agent systems with event-triggered consensus protocol. For the linear systems subject to input saturation, both the dynamic event-triggering and self-triggering mechanism are proposed and the stability of the closed-loop systems is discussed with anti-windup compensation in [233].

1.4 OUTLINE OF THIS BOOK

In this book, we aim to discuss the problems of networked and event-triggered control in CPSs. In order to cope with network communication uncertainties, such as network communication delay, data dropout, and signal quantization, the stabilization and delay compensation mechanisms in CPSs are presented in Section I of this book. In Section II, both the event-triggered and self-triggered control schemes are derived for the CPSs to reduce the communication loads. Meanwhile, to compensate both the network communication uncertainties and external disturbance on CPSs, the event-triggered disturbance rejection control methods are also proposed in this section. Then, specific research contents in Section I are as follows:

- In Chapter 2, by considering the random characteristics of network communication delay and data dropout, the stochastic modeling of a class of CPS is developed based on the probability distribution of the delay and data dropout. Different from existing modeling approaches, the presented CPS model incorporates the phenomena of data dropout and communication delay under a unified probabilistic framework. Based on the developed CPS model, stochastic stability and stabilization are further considered. Both state and output feedback control schemes are investigated when the probabilities are known, partially known or completely unknown, respectively. Compared with existing modeling approaches, the established conditions are necessary and sufficient, and have less conservatism.

- In Chapter 3, the observer-based output feedback control problem for discrete-time linear systems with quantized inputs is considered. The basic control configuration is to reconstruct the state by using an observer, then sending the quantized control input to the controlled plant. The polytopic idea is employed to derive the quadratic stabilizability condition, then a single step design procedure is proposed to design the controller and observer gains. To solve out the control and observer gains, the necessary and sufficient condition is proposed to guarantee the asymptotic stability of the closed-loop systems, and a quantization-density-dependent necessary and sufficient condition is proposed. Then, both the full and reduced order observer-based quantized feedback controller are designed to overcome the drawbacks of the two-step LMI approach often encountered in the literature.

- Chapter 4 addresses the problem of networked delay compensation control and stability analysis for CPSs with time-varying network communication delay. As for networked delay compensation control scheme, two important problems should be solved, one is how to construct the control input predictions (CIPs), and the other is how to analyze the stability of the closed-loop networked control system. In this chapter, by taking the full advantages of the packet-based transmission in CPSs, a state-based networked delay compensation control approach is proposed to actively compensate the network communication delay. Moreover, the stability analysis is also performed by applying the average dwell time technique.

- In Chapter 5, the design problem of static output feedback delay compensation controller for CPSs with random network delay is considered, and the delay compensation control approach is proposed to actively compensate the delay under Markovian jump linear system framework. Different from previously reported delay compensation control approaches for CPSs, an output feedback strategy is used to generate the CIPs, and the necessary and sufficient condition is proposed to guarantee the stability analysis of networked closed-loop system. Furthermore, the controller design problem is solved by using the singular Markovian jump system approach.

- Chapter 6 is concerned with the network delay compensation control problem for nonlinear CPSs, and the network delay compensation approaches are proposed to actively compensate the network communication delay under the fuzzy control framework. The nonlinear plant is represented by a Takagi-Sugeno (T-S) fuzzy model, and the CIPs are constructed based on parallel distributed compensation technique. Both state and output feedback fuzzy delay compensation controllers are designed for the considered nonlinear plants.

- In Chapter 7, the problem of networked output tracking control is investigated by considering the delay compensations in both the sensor-to-controller and controller-to-actuator channels in CPSs. The delayed output measurements are treated as a special output disturbance, and the sensor-to-controller channel delay is compensated with the aid of an extended functional state observer. For the delay in the controller-to-actuator channel, the buffer and packet-based delay compensation approaches are presented, respectively. Then, the stability analysis is performed for the networked closed-loop systems.

The next section of this book is to address event-triggered control approaches in CPSs.

- Chapter 8 concerns the problem of observer-based event-triggered output feedback control for linear systems. Contrary to normal sampled-data control systems, where the controller is updated periodically, in this chapter, it is updated only when an "event" happens, and a typical event is defined as some error signals exceeding a given event threshold. Both continuous and discrete updating instants scheduler (UIS) cases are considered. It is shown that even with the significantly reduced updating frequency of the controller, the global uniform ultimate boundedness of the states of the event-triggered closed-loop systems can also be guaranteed.

- In Chapter 9, the self-triggered output feedback control problem is addressed for discrete-time systems, and the UIS is implemented to determine when the controller is updated. For both the full- and reduced-order observer cases, the updating instants are determined, respectively, where only the information of the estimated state at the current updating instant is required to obtain the next updating instant. It is shown that, with the proposed self-triggered control schemes, not only the updating frequency is significantly reduced, but also the uniform ultimate boundedness of the closed-loop system is guaranteed.

- In Chapter 10, the dynamic output feedback (DOF) controllers are designed for systems with event-triggered control inputs, which can significantly save the data transmissions and in the meanwhile maintain the system performance at a satisfactory level. The dynamic output feedback controllers are designed for systems with either continuous or sampled output measurements, and the control signals are transmitted according to a predefined UIS. It is shown that, with the proposed event-triggered control scheme, the global uniform ultimate boundedness of the closed-loop systems is guaranteed and the inter-event interval is lower bounded by a positive scalar.

- In Chapter 11, the co-design of event-triggered control and quantized control techniques is considered, and new quantized event-triggered controller is designed for both the fixed and relative event threshold cases. First, similar to the predictive vector quantizer, the encoder and decoder, the UIS, and event-based quantizer are designed for event-triggered control system, and the stability analysis problems are addressed for the quantized event-triggered control systems with fixed and relative event thresholds, respectively. For the fixed event threshold case, the range of quantization length is obtained and then the states of closed-loop system are shown to be globally uniformly ultimately bounded (UUB). For the relative event threshold case, the modified quantizer and the UIS are proposed, and the globally asymptotic stability of the closed-loop system is guaranteed. Furthermore, the existence on the minimum inter-event interval to exclude the Zeno behavior is shown for both the two cases.

- Chapter 12 proposes the event-triggered disturbance rejection control approaches for discrete-time systems with external disturbance. Due to the existence of unknown external disturbance, it is imprecise to directly use the system states to determine the UIS and to design the feedback controllers. To address this problem, the states and disturbance are estimated by using an extended state observer (ESO), and then these estimates are utilized to establish the UIS and controller. Then, the reduced-order ESO case is further considered. Furthermore, the stability and disturbance rejection properties are also discussed. It is shown that the use of the proposed event-triggered control strategy can lead to not only a substantial decrease in the updating frequency of the controllers but also an effective rejection of the disturbance from the system output channel.

Above all, both the networked and event-triggered control approaches are well investigated for a series of CPSs, and all theoretical results of this book are verified by experiments or numerical simulations.

I

Networked Control Approaches in CPSs

CHAPTER 2

Stochastic Stability and Stabilization of CPSs with Random Delay

2.1 INTRODUCTION

In CPSs, owing to the communication constraints, the network communication delay and data dropout are two general issues should be considered, and the former is usually caused by limited bit rate of the communication channels, a node waiting to send out a packet via a busy channel, or signal processing and propagation, and the later is usually caused by the unavoidable errors or losses in the transmission [71, 148, 56]. In general, network communication delay and data dropout are random phenomenona, and will degrade the control performance of the CPSs. Therefore, it is important to investigate the stochastic stability and stabilization problems for CPSs with simultaneous random network communication delay and data dropout [114, 122, 22, 190]. It should be noted that, in some situations, such as those systems connected over a wireless network, the observation of communication delay and/or data packet dropout also include the probability distribution of their values [190, 44]. Therefore, it is realistic and accurate to take into account the occurrence probabilities of the delay values and data dropout in CPSs. In [190], the stability analysis and stabilization controller design are considered for continuous-time system with a stochastic input delay taking values in some intervals, which obeys a probability distribution. In [44], by taking into account the occurrence probabilities of the delay values, the problem of stabilization for network-based control systems is investigated.

In this chapter, by considering the random characteristics of network communication delay and data dropout, the modeling approach is developed based on the probability distribution of the delay and data dropout. Different from existing modeling approaches, the presented model incorporates the phenomena of data dropout and communication delay under a unified probabilistic framework. Based on the developed model, the stochastic stability and stabilization are further considered. Both state and output feedback control schemes are investigated when the probabilities are known, partially known or completely unknown, respectively. Compared with existing modeling approaches, the established conditions are necessary and sufficient, and have less conservatism. Finally, three numerical examples are provided to demonstrate the effectiveness and reduced conservatism of the proposed methods.

2.2 SYSTEM DESCRIPTION

The dynamics of the remote-controlled plant in the CPS is given by the following multi-input-multi-output (MIMO) discrete-time system:

$$x(k+1) = Ax(k) + Bu(k)$$
$$y(k) = Cx(k) \tag{2.1}$$

where $x(k) \in \mathbb{R}^n$ is the state vector; $u(k) \in \mathbb{R}^m$ is the control input; $y(k) \in \mathbb{R}^q$ is the output; and A, B, C are system matrices with appropriate dimensions. Also, the following assumption is made.

Assumption 2.1. *The pair (A, B) is completely controllable, and the pair (A, C) is completely observable.*

The sensor is time-triggered. At each time step k, the system state packet $\vec{x}(k) = [x^T(k), x^T(k-1)]^T$ or output packet $\vec{y}(k) = [y^T(k), y^T(k-1)]^T$ with its time stamp (that is, the time the plant state is sampled) are encapsulated into a packet and sent to the controller via the network.

In practice, due to the limited bit rate of the communication channel and unavoidable errors or losses in the transmission, the data packets in CPSs usually suffer from time delays and data packet dropout during network transmissions. Let d_k^{sc} and d_k^{ca} be the network communication delay in the sensor-to-controller and controller-to-actuator channel, respectively, and let $d_k = d_k^{sc} + d_k^{ca}$ be the round trip delay at time step k. Also, the following assumptions are made:

Assumption 2.2. *At each time step k, the round trip delay d_k is random and bounded, that is, $1 \leq d_m \leq d_k \leq d_M$, where d_m and d_M are positive integers.*

Assumption 2.3. *The data dropout occurs randomly, and the number of consecutive data dropouts is bounded.*

To characterize the random network communication delay and data dropout, the following possible basic assumption is considered in this chapter.

Assumption 2.4. *The network communication delay and data dropout are assumed to be i.i.d. (independent and identically-distributed), with the probabilities given by*

$$\Pr\{d_k = i\} = p_i, \ i \in \bar{\Pi}, \tag{2.2}$$

where $\bar{\Pi} \triangleq \Pi \cup \{\infty\}$ with $\Pi \triangleq \{d_m, d_m + 1, \ldots, d_M\}$, $p_i \geq 0$ and $\sum_{i \in \bar{\Pi}} p_i = 1$, and $i = \infty$ corresponds to the data packet dropout.

At time step k, it is assumed the state packet $\vec{x}(k - d_k^{sc})$ or the output packet $\vec{y}(k - d_k^{sc})$ is the received at the controller side. The networked controller is selected as state-based PD controller:

$$u(k) = K_P x(k - d_k^{sc}) + K_D(x(k - d_k^{sc}) - x(k - d_k^{sc} - 1)), \qquad (2.3)$$

or output-based PD controller:

$$u(k) = \tilde{K}_P y(k - d_k^{sc}) + \tilde{K}_D(y(k - d_k^{sc}) - y(k - d_k^{sc} - 1)), \qquad (2.4)$$

where K_P, K_D, $\tilde{K}_P$, and $\tilde{K}_D$ are the controller gains to be designed later. The controller is event-triggered. As long as there is a packet reaching the controller across the network from the sensor, the controller calculates the control signal immediately. After finishing the calculations, the new control signal and the time stamp of the used plant state or output are encapsulated into a packet and sent to the actuator via network.

The buffer has the capability to compare the time stamps of the arrived control signal and the one stored in it, and will be updated only when the arrived control input is more recent than the existing one. The actuator is event-triggered. The control signal is applied to the plant as soon as the data packet arrives at the actuator node.

With the above observations in mind, the networked control system can be rewritten as a linear discrete-time system with input delay d_k. For state-feedback case, the closed-loop system can be written as

$$x(k+1) = Ax(k) + B(K_P + K_D)x(k - d_k) - BK_D x(k - d_k - 1) \qquad (2.5)$$

and for output feedback case, the closed-loop system can be written as

$$x(k+1) = Ax(k) + B(\tilde{K}_P + \tilde{K}_D)y(k - d_k) - B\tilde{K}_D y(k - d_k - 1) \qquad (2.6)$$

where d_k is the round trip delay at time step k.

2.3 MODELING OF CPS

In a network environment, the signal transmitted in network medium will suffer from random communication delays and data dropouts. It is worth noting that, for each time instant k, the random communication delay d_k can only take one value from the set Π randomly. To model an CPS, the following augmented state vector is introduced:

$$\bar{x}(k) = \begin{bmatrix} x^T(k) & x^T(k - 1) & \cdots & x^T(k - d_M - 1) \end{bmatrix}^T.$$

Thus, when there is a network delay d_k taking value in Π randomly, system (2.5) can be augmented as

$$\bar{x}(k+1) = \bar{A}_i \bar{x}(k) \qquad (2.7)$$

where

$$\bar{A}_i = \left[\begin{array}{ccccc} A & 0_{n \times (i-1)n} & B(K_P + K_D) & -BK_D & 0_{n \times (d_M - i)n} \\ \hline I_{(d_M+1)n \times d_M n} & & I_{(d_M+1)n \times n} & & 0_{(d_M+1)n \times n} \end{array} \right], i \in \Pi.$$

Moreover, when the data dropout phenomenon occurs (it corresponds to the network delay being ∞), system (2.5) becomes an open-loop system and system matrix in (2.7) can be given as

$$\bar{A}_\infty = \left[\begin{array}{c|c} A & 0_{n \times (d_M+1)n} \\ \hline I_{(d_M+1)n \times (d_M+1)n} & 0_{(d_M+1)n \times n} \end{array} \right].$$

Thus, at each time instant k, the networked control system (2.5) with random communication delays and data dropouts can be modeled as

$$\bar{x}(k+1) = \bar{A}_{\sigma_k} \bar{x}(k), \tag{2.8}$$

where $\{\sigma_k\} \in \bar{\Pi}$ is i.i.d. with probability distribution $\{p_{d_m}, \ldots, p_{d_M}, p_\infty\}$.

Before proceeding further, the following definition is presented.

Definition 2.1. *System (2.8) is said to be stochastically stable in the mean square for any initial condition $\bar{x}_0$ if there exists a finite $W > 0$ such that*

$$\mathbb{E} \left\{ \sum_{k=0}^{\infty} \|\bar{x}(k)\|^2 \, \Big| \, \bar{x}_0 \right\} < \bar{x}_0^T W \bar{x}_0.$$

The objective of this chapter is, under Assumption 2.4, to determine the PD gain matrices K_P, K_D and $\tilde{K}_P$, $\tilde{K}_D$ of the controllers in (2.3) and (2.4), respectively, such that the closed-loop system (2.8) is stochastically stable in the mean square for all admissible random network communication delays and data packet dropouts.

2.4 STOCHASTIC STABILITY AND STABILIZATION

The following result on stochastic stability of system (2.8) is presented firstly. Based on the Corollary 2.2 in [29] or Corollary 3.26 in [17], the following result can be obtained immediately.

Theorem 2.1. *System (2.8) is stochastically stable in the mean square if and only if there exists a matrix $P > 0$ satisfying the following matrix inequality,*

$$\sum_{i \in \bar{\Pi}} p_i \bar{A}_i^T P \bar{A}_i - P < 0. \tag{2.9}$$

Remark 2.1. *It should be pointed out that, the stability condition in Theorem 2.1 is established based on the augmented system (2.7), and the dimension of $\bar{A}_i$ is directly related to the upper bound d_M, which implies that the dimension of the augmented system may be larger if the time-delay is larger.*

In Theorem 2.1, the probability p_i is assumed to be completely known, however, in practice, it may be more reasonable to assume that p_i belongs to a interval, say $p_i \in [\underline{p}_i, \bar{p}_i]$, where $0 \leq \underline{p}_i$ and $\bar{p}_i \leq 1$. For simplicity, we denote $\tau_i = \frac{\bar{p}_i + \underline{p}_i}{2}$ and $\delta_i = \frac{\bar{p}_i - \underline{p}_i}{2}$, thus we have $p_i \in [\tau_i - \delta_i, \tau_i + \delta_i]$. Then, we have the following result.

Theorem 2.2. *Assume that $p_i \in [\underline{p}_i, \bar{p}_i]$, $i \in \bar{\Pi}$, are partially known. System (2.8) is stochastically stable in the mean square, if and only if there exists a matrix $P > 0$ satisfying the following matrix inequalities,*

$$\mathcal{A}^T \Delta_p^{(i)} \mathcal{P} \mathcal{A} - P < 0, \ i = 1, 2, \dots, 2^\nu, \tag{2.10}$$

where $\nu = d_M - d_m + 2$, $\Delta_p^{(i)}$, $i = 1, 2, \dots, 2^\nu$, represents the 2^ν distinct matrices given in terms of $\mathrm{diag}\{\tau_{d_m} \pm \delta_{d_m}, \tau_{d_m+1} \pm \delta_{d_m+1}, \dots, \tau_{d_M} \pm \delta_{d_M}, \tau_\infty \pm \delta_\infty\}$, and

$$\mathcal{A} = \begin{bmatrix} \bar{A}_{d_m}^T & \bar{A}_{d_m+1}^T & \cdots & \bar{A}_{d_M}^T & \bar{A}_\infty^T \end{bmatrix}^T,$$
$$\mathcal{P} = \mathrm{diag}\{P, P, \dots, P, P\} \in \mathbb{R}^{\nu n \times \nu n}.$$

Proof. It should be noted that (2.9) can be rewritten into the following form:

$$\mathcal{A}^T \Delta_p \mathcal{P} \mathcal{A} - P < 0,$$

where $\Delta_p = \mathrm{diag}\{p_{d_m}, \dots, p_{d_M}, p_\infty\}$. Noticing that $p_i \in [\tau_i - \delta_i, \tau_i + \delta_i]$, thus Δ_p can be rewritten into in the following polytopic form:

$$\Delta_p = \sum_{i=1}^{2^\nu} \alpha_i \Delta_p^{(i)}, \ \sum_{i=1}^{2^\nu} \alpha_i = 1, \ \alpha_i \geq 0. \tag{2.11}$$

Next, we will show (2.9) is equivalent to (2.10). If (2.9) holds, by taking $p_i = \tau_i - \delta_i$ or $\tau_i + \delta_i$ for $i = 1, 2, \dots, 2^\nu$, respectively, we obtain (2.10) immediately. If (2.10) holds and note that (2.11), it can be verified that

$$\sum_{i \in \bar{\Pi}} p_i \bar{A}_i^T P \bar{A}_i - P = \sum_{i=1}^{2^\nu} \alpha_i \left\{ \mathcal{A}^T \Delta_p^{(i)} \mathcal{P} \mathcal{A} - P \right\} < 0,$$

which completes the proof. □

Furthermore, if we do not have any information on probabilities p_i, $i \in \bar{\Pi}$, the following corollary can be used to test the stability of system (2.8).

Corollary 2.1. *Assume p_i, $i \in \bar{\Pi}$ are unknown. System (2.8) is stochastically stable in the mean square, if there exists a matrix $P > 0$ satisfying*

$$\bar{A}_i^T P \bar{A}_i - P < 0, \ i \in \bar{\Pi}. \tag{2.12}$$

Proof. Consider of $\sum_{i \in \bar{\Pi}} p_i = 1$, (2.9) can be rewritten as

$$\sum_{i \in \bar{\Pi}} p_i \left(\bar{A}_i^T P \bar{A}_i - P \right) < 0. \tag{2.13}$$

Note that $p_i \geq 0$, $i \in \bar{\Pi}$ and (2.13), (2.12) implies that (2.10) holds. □

Remark 2.2. *It should be pointed out that if $p_i, i = 1, 2, \ldots, \nu$ are completely unknown, system (2.8) becomes a switched linear system under arbitrary switching [172], and condition (2.12) is a sufficient condition to guarantee that the switched linear system is globally asymptotically stable under a discrete-time framework.*

With the stability analysis results, we further consider the problem of controller design in this subsection.

Theorem 2.3. *System (2.5) is stochastically stable in the mean square via state feedback, if and only if there exist matrices $P > 0$, $Q > 0$ and PD gain matrices K_P and K_D satisfying the following conditions:*

$$\begin{bmatrix} -P & \mathcal{A}_p^T \\ \mathcal{A}_p & -\mathcal{Q} \end{bmatrix} < 0, \tag{2.14}$$

$$PQ = I, \tag{2.15}$$

where

$$\mathcal{A}_p = \begin{bmatrix} \sqrt{p_{d_m}} \bar{A}_{d_m}^T & \sqrt{p_{d_m+1}} \bar{A}_{d_m+1}^T & \cdots & \sqrt{p_{d_M}} \bar{A}_{d_M}^T & \sqrt{p_\infty} \bar{A}_\infty^T \end{bmatrix}^T,$$

$$\mathcal{Q} = \mathrm{diag}\{Q, Q, \ldots, Q, Q\} \in \mathbb{R}^{\nu n \times \nu n}.$$

Proof. According to Schur complement lemma, (2.9) is equivalent to

$$\begin{bmatrix} -P & \mathcal{A}_p^T \\ \mathcal{A}_p & -\mathcal{P}^{-1} \end{bmatrix} < 0, \tag{2.16}$$

where $\mathcal{P} = \mathrm{diag}\{P, P, \ldots, P, P\} \in \mathbb{R}^{(\nu-1)n \times (\nu-1)n}$. Defining $Q = P^{-1}$ in (2.16) gives (2.14) and (2.15). This completes the proof. □

Because of the conditions in Theorem 2.3 are no longer LMIs owing to (2.15), we cannot use a convex optimization algorithm to design the PD controller gain K_P and K_D. However, we can resort to the cone complementarity linearization (CCL) algorithm to convert the original problem into the following nonlinear minimization problem:

$$\begin{aligned} \text{Minimize} \quad & \text{trace}(PQ) \\ \text{subject to} \quad & (2.14) \text{ and,} \end{aligned}$$

$$P > 0, \ Q > 0, \tag{2.17}$$

$$\begin{bmatrix} P & I \\ I & Q \end{bmatrix} \geq 0. \tag{2.18}$$

Then, for given probabilities $p_i, i \in \bar{\Pi}$, the controller gain K can be determined by the following algorithm:

Algorithm 2.1 Controller Design Algorithm (Known Probabilities Case):

1: Find a feasible set of P_0 and Q_0 satisfying (2.14), (2.17), and (2.18). Set $k = 0$.

2: Solve the following LMI problem for the variables P, Q, K_P, and K_D:

$$\text{Minimize} \quad \text{trace}\{P_k Q + P Q_k\}$$
$$\text{subject to} \quad (2.14), (2.17), \text{ and } (2.18)$$

Set $P_{k+1} = P$, $Q_{k+1} = Q$.

3: With the obtained K in Step 2, if (2.9) are feasible for the variable P, then exit. If (2.9) is infeasible within a specified number of iterations, then stop. Otherwise, set $k = k + 1$ and go to Step 2.

Furthermore, if only partially or completely unknown information of probabilities p_i can be obtained, one can use the following results to design networked state-feedback controller.

Theorem 2.4. *System (2.5) with $p_i \in [\underline{p}_i, \overline{p}_i]$, $i \in \Pi$, is stochastically stable in the mean square via state feedback, if and only if there exist matrices $P > 0$, $Q > 0$ and PD gain matrices K_P and K_D satisfying the following conditions:*

$$\begin{bmatrix} -P & \mathcal{A}^T \bar{\Delta}_p^{(i)} \\ \bar{\Delta}_p^{(i)} \mathcal{A} & -\mathcal{Q} \end{bmatrix} < 0, \tag{2.19}$$

$$PQ = I, \tag{2.20}$$

where $\bar{\Delta}_p^{(i)}$, $i = 1, 2, \ldots, 2^\nu$, represents the 2^ν distinct matrices given in terms of

$$diag\left\{ \sqrt{\tau_{d_m} \pm \delta_{d_m}}, \sqrt{\tau_{d_m+1} \pm \delta_{d_m+1}}, \ldots, \sqrt{\tau_{d_M} \pm \delta_{d_M}}, \sqrt{\tau_\infty \pm \delta_\infty} \right\}.$$

The results can be obtained by using Theorem 2.2, and following a similar line of Theorem 2.3, we can obtain Corollary 2.2.

Corollary 2.2. *System (2.5) with unknown probabilities p_i, $i \in \bar{\Pi}$, is stochastically stable in the mean square via state feedback, if there exist matrices $P > 0$, $Q > 0$ and PD gain matrices K_P and K_D satisfying the following conditions:*

$$\bar{A}_\infty^T P \bar{A}_\infty - P < 0, \tag{2.21}$$

$$\begin{bmatrix} -P & \bar{A}_i^T \\ \bar{A}_i & -Q \end{bmatrix} < 0, \ i \in \Pi, \tag{2.22}$$

$$PQ = I. \tag{2.23}$$

It should be pointed out that, in practical applications, the state variables are often not all available. In this situation, it is necessary to design an output feedback controller. The results of the stability analysis and controller design for the state-feedback case are now employed to design static output feedback stabilization controllers for system (2.1).

Assume that the static output feedback control law is given by $u(k) = \tilde{K}y(k)$ where $\tilde{K} \in \mathbb{R}^{m \times q}$ is the controller gain. Applying $u(k) = \tilde{K}y(k)$ to system (2.5) yields the following closed-loop system $x(k+1) = Ax(k) + B\tilde{K}Cx(k-d_k)$. Following a similar line, we obtain the augmented closed-loop system:

$$\bar{x}(k+1) = \tilde{A}_{\sigma_k}\bar{x}(k), \ \sigma_k \in \bar{\Pi} \tag{2.24}$$

where

$$\tilde{A}_{\sigma_k} = \left[\begin{array}{ccccc} A & 0_{n \times (\sigma_k-1)n} & B(\tilde{K}_P + \tilde{K}_D)C & -B\tilde{K}_DC & 0_{n \times (d_M-\sigma_k)n} \\ \hline & I_{d_Mn \times d_Mn} & & 0_{d_Mn \times n} & \end{array} \right], \ \sigma_k \in \Pi,$$

and

$$\tilde{A}_\infty = \left[\begin{array}{c|c} A & 0_{n \times d_Mn} \\ \hline I_{d_Mn \times d_Mn} & 0_{d_Mn \times n} \end{array} \right], \ \sigma_k = \infty.$$

Then, output feedback controllers can then be designed similarly when the probabilities p_i is completely known, partially known or completely unknown, respectively.

2.5 NUMERICAL SIMULATION

In this section, three examples are provided to demonstrate the effectiveness of the proposed results in this chapter.

Example 2.1. *Consider the following aircraft model borrowed from [49]:*

$$\dot{x}(t) = Gx(t) + Hu(t)$$
$$y(t) = Cx(t)$$

where

$$G = \begin{bmatrix} -2.98 & 0.93 & 0 & -0.0340 \\ -0.99 & -0.21 & 0.035 & -0.0011 \\ 0 & 0 & 0 & 1 \\ 0.39 & -5.555 & 0 & -1.89 \end{bmatrix},$$

$$H = \begin{bmatrix} -0.032 \\ 0 \\ 0 \\ -0.16 \end{bmatrix}, \ C = \begin{bmatrix} 0 & 0 & 1 & 0 \\ 0 & 0 & 0 & 1 \end{bmatrix}.$$

Taking the sampling period as $T = 0.05s$, we obtain the discrete system (2.1) with

$$A = \begin{bmatrix} 0.8605 & 0.0432 & 0.0000 & -0.0015 \\ -0.0457 & 0.9885 & 0.0017 & 0.0000 \\ 0.0006 & -0.0067 & 1.0000 & 0.0477 \\ 0.0236 & -0.2631 & -0.0002 & 0.9098 \end{bmatrix}, \ B = \begin{bmatrix} -0.0015 \\ 0.0000 \\ -0.0002 \\ -0.0077 \end{bmatrix}.$$

In the following, the networked state and output feedback controllers in the form of (2.3) and (2.4) are designed such that the closed-loop system is stochastically stable in the mean square for all admissible random network communication delays and data packet dropouts.

Now, we assume the lower and upper bound of network communication delay is $d_m = 1$ and $d_M = 3$, respectively, and random network delays and data dropouts satisfy the probability distribution shown in Table 2.1.

Table 2.1 Probability distribution of random network delays and data dropouts

d_k	$d_k = 1$	$d_k = 2$	$d_k = 3$	Data Dropouts
Probability	0.2	0.3	0.3	0.2

By resorting to the controller design algorithm, the state-based PD controller gain matrices K_P and K_D can be computed as

$$K_P = \begin{bmatrix} -3.3555 & -1.5670 & 7.2682 & 34.3393 \end{bmatrix},$$

$$K_D = \begin{bmatrix} 26.5778 & -18.4867 & 9.6873 & 11.4528 \end{bmatrix}.$$

Simulation results are shown in Figure 2.1, where the initial condition is assumed to be $[0, 0, 1, -0.5]^T$. It can be seen that the state variables converge to zero under the random network delay and data dropouts, which shows the effectiveness of the controller design approach.

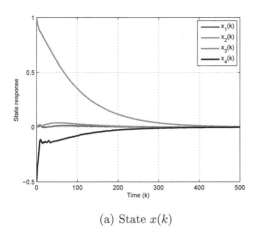

(a) State $x(k)$

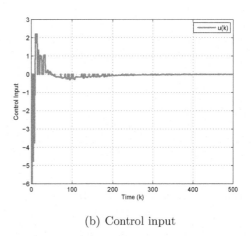

(b) Control input

Figure 2.1 State and control input signals of aircraft system (state-feedback case)

Next, we further consider the design of networked output feedback controller for system (2.1). Under the same probability distribution, according to the controller design algorithm, the output-based PD controller gain matrices $\tilde{K}_P$ and $\tilde{K}_D$ can be given as

$$\tilde{K}_P = \begin{bmatrix} 5.4707 & 33.7438 \end{bmatrix}, \tilde{K}_D = \begin{bmatrix} 0.8952 & 16.2864 \end{bmatrix}.$$

The state responses are depicted in Figure 2.2, from which we can see that the state components converge to zero under the random network delay and data dropouts.

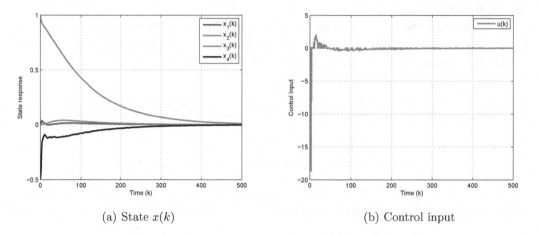

(a) State $x(k)$ (b) Control input

Figure 2.2 State and control input signals of aircraft system (output-feedback case)

Next, we use the following example to demonstrate the effectiveness of our method and its advantage over existing ones.

Example 2.2. *Consider the following discrete-time system with time-varying delay:*

$$x(k+1) = Ax(k) + A_d x(k - d_k), \qquad (2.25)$$

where A and A_d are constant matrices with appropriate dimensions, d_k is a time-varying state delay satisfying $1 \le d_m \le d_k \le d_M$.

It can be seen that the stability analysis approach proposed in Section 2.4 can also be used to analyze the stability of system (2.25), and we assume system (2.25) with the following parameters:

$$A = \begin{bmatrix} 0.8 & 0 \\ 0.05 & 0.9 \end{bmatrix}, \ A_d = \begin{bmatrix} -0.1 & 0 \\ -0.2 & -0.1 \end{bmatrix},$$

which has been considered in [43, 38, 66, 110].

Now, we assume the lower bound d_m is given, and we are interested in the upper delay bound d_M below which the above system (2.25) is asymptotically stable for all $d_m \le d_k \le d_m$. A detailed comparison is given in Table 2.2, where the achieved upper bounds of delay d_k in system (2.25) are listed for their respective lower bounds. Clearly, the method presented in this chapter is significantly better than those in existing results.

Furthermore, in [110], a delay-partitioning approach is used to analyze the stability of system (2.25). Although delay-partitioning approach can reduce the conservatism distinctly over the approaches proposed in [43, 38, 66], there is an assumption on delay lower bound d_m, that is $d_m = \tau m$ where τ and m are integers, which implies

Table 2.2 Allowable upper bounds of d_M for various d_m

d_m	2	4	6	10	12
[43]	7	8	9	12	13
[38]	13	13	14	15	17
[66]	17	17	18	20	21
[204]	19	20	21	23	24
Corollary 2.1	22	22	23	24	25

that the delay-partitioning approach is available for some pairs of (τ, m). For example, $d_m = 4$, the delay-partitioning approach is valid for two cases, $\tau = 2$, $m = 2$ and $\tau = 4$, $m = 1$. However, this limitation is not suitable for our approach, and our approach will lead less conservative results over [110], see Table 2.3 for more details. It can be seen that our approach is better than the delay-partitioning approach.

Table 2.3 Comparison with delay-partitioning approach

d_m	4	12	16
[110]	17 $(m = 2, \tau = 2)$	21 $(m = 1, \tau = 12)$	24 $(m = 1, 2, \tau = 16, 8)$
[110]	18 $(m = 4, \tau = 1)$	22 $(m = 2, 3, 4, 6, 12,$ $\tau = 6, 4, 3, 2, 1)$	25 $(m = 4, 8, 16, \tau = 4, 2, 1)$
Corollary 2.1	22	25	27

2.6 SUMMARY

In this chapter, the problem of modeling, stability and stabilization have been considered for CPSs by carefully considering the stochastic characteristics of both random network communication delay and data dropout. By using the probability distribution of delay and data dropout, the problems of stability analysis and state and output feedback stabilization have been further considered when the probabilities are known, partially known or completely unknown, respectively. Finally, three numerical examples are provided to demonstrate the effectiveness and reduced conservatism of the proposed method.

CHAPTER 3

Observer-Based Output-Feedback Stabilization for CPSs with Quantized Inputs

3.1 INTRODUCTION

In CPSs, because of the limited bandwidth of the network, signals are always quantized before being communicated. In recent years, signal quantization in feedback control system has received a great deal of attention, which is mainly due to the widespread use in control systems of digital computers that employ finite-precision arithmetic [40]. In recent studies, quantizers are usually regarded as information coders, and the most important problem is how much information is required to be communicated by the quantizer in order to achieve a certain performance objective of the closed-loop system. By following this line, signal quantization has been investigated by many researchers for both linear systems [9, 25, 35, 61] and nonlinear systems [97, 62]. The existing quantization policies can be categorized into two types, static and dynamic quantizations. Static quantization policy assumes that data quantization is dependent on the data at current time only, which leads to relatively simple structures for the coding/decoding schemes, see [19, 27, 25, 35, 40, 125] for example. On the other hand, dynamic quantizers are always time-varying and have memory, and it can outperform the conventional static quantizers. Design approaches of dynamic quantizers can be found in [9, 95, 111], and the references therein. More recently, there are applications of these quantization policies to networked control systems, for instance, [188, 41, 120, 12, 115, 147, 130], and the references therein.

In this chapter, by using a polytopic approach, we consider observer-based output feedback control for discrete-time linear systems with quantized inputs. The basic control configuration is to reconstruct the state by using a observer, then sending the quantized control input to the controlled plant. We firstly employ the polytopic idea to derive the quadratic stabilizability condition, then we present a single step

DOI: 10.1201/9781003260882-3

procedure to design the controller and observer gains. To solve the controller and observer gains, the necessary and sufficient condition is proposed to guarantee the asymptotic stability of the closed-loop systems. Different from [35, 40], the quantization-density-dependent necessary and sufficient condition is derived, and the design approach is proposed to design both full- and reduced-order observer-based quantized feedback controllers, which overcomes the drawbacks of the two-step LMI approach often encountered in the literature. Finally, illustrative example is given to demonstrate the effectiveness and the benefits of the proposed method.

3.2 SYSTEM DESCRIPTION

Consider the following discrete-time system:

$$x(k+1) = Ax(k) + Bu(k)$$
$$y(k) = Cx(k) \tag{3.1}$$

where $A \in \mathbb{R}^{n \times n}$, $B \in \mathbb{R}^{n \times m}$, $C \in \mathbb{R}^{q \times n}$, $x(k)$ is the state and $u(k)$ is the control input, and the pairs (A, B) and (A, C) are assumed to satisfy Assumption 2.1.

The quantizer is denoted as $q[\cdot] = [q_1[\cdot], q_2[\cdot], \ldots, q_m[\cdot]]^T$, which is assumed to be symmetric, that is $q_j[-\varepsilon] = -q_j[\varepsilon], j = 1, 2, \ldots, m$. In this chapter, the quantizer is selected as a logarithmic one, shown in Figure 3.1, and for each $q_j[\cdot]$, the quantization

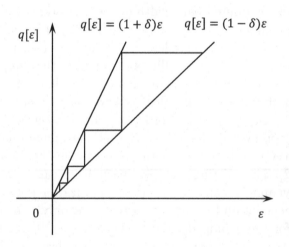

Figure 3.1 Logarithmic quantizer

levels given by

$$\mathcal{V}_j = \{\pm\mu_i^{(j)} : \mu_i^{(j)} = \rho_j^i \mu_0^{(j)}, i = 0, \pm 1, \pm 2, \ldots\} \cup \{0\}, \ \mu_0 > 0$$

where $\rho_j \in (0, 1)$, $j = 1, 2, \ldots, m$, are scalars. According to [25, 35], the associated quantizer is defined as follows:

$$q_j[\varepsilon] = \begin{cases} \rho_j^i \mu_0^{(j)} & \text{if } \frac{1}{1+\delta_j}\rho_j^i \mu_0^{(j)} < \varepsilon \le \frac{1}{1-\delta_j}\rho_j^i \mu_0^{(j)}, \ \varepsilon > 0, \\ 0 & \text{if } \varepsilon = 0, \\ -q_j[-\varepsilon] & \text{if } \varepsilon < 0, \end{cases} \tag{3.2}$$

where $\delta_j = \frac{1-\rho_j}{1+\rho_j}$.

For quantizer (3.2), we recall the following definition.

Definition 3.1. [25, 35] *Denote by $\#g[\epsilon]$ the number of quantization levels in the interval $[\epsilon, 1/\epsilon]$. Then, the density of quantizer $q[\cdot]$ is defined as follows:*

$$\eta_q = \limsup_{\epsilon \to 0} \frac{\#g[\epsilon]}{-\ln \epsilon}.$$

With the above definition, it is not difficult to verify that the density of logarithmic quantizer (3.2) is $\eta_q = \frac{2}{\ln\frac{1}{\rho}}$, which means that the smaller the ρ, the smaller the η_q. Thus, we will abuse the terminology by calling ρ (instead of η_q) the quantization density.

In this chapter, we consider the observer-based quantized feedback control, as depicted in Figure 3.2, and the observer-based quantized feedback controller is given by

$$\begin{aligned} \hat{x}(k+1) &= A\hat{x}(k) + Bu(k) + L(y(k) - \hat{y}(k)) \\ \hat{y}(k) &= C\hat{x}(k) \\ v(k) &= K\hat{x}(k) \\ u(k) &= q[v(k)] \end{aligned} \tag{3.3}$$

where $\hat{x}(k)$ is the state of observer, $q[\cdot]$ is the logarithmic quantizer (3.2), and K and L are the controller and observer gains, respectively.

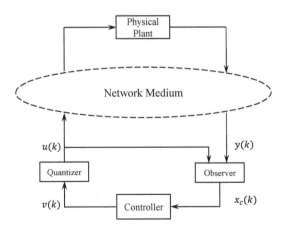

Figure 3.2 Block diagram of observer-based quantized feedback control

The objective of this chapter is to design the observer-based quantized controller for system (3.1), that is, to determine the gain matrices K and L in (3.3), such that the closed-loop system is asymptotically stable for admissible signal quantization errors.

3.3 FULL-ORDER OBSERVER-BASED QUANTIZED CONTROL

To obtain the closed-loop system, we define the quantization error by

$$e(k) = u(k) - v(k) = q[v(k)] - v(k), \tag{3.4}$$

where $v(k) = K\hat{x}(k)$. It is not difficult to verify that $(1 - \delta_i)v_i(k) \le q_i[v_i(k)] \le (1 + \delta_i)v_i(k)$ and $-\delta_i v_i(k) \le q_i[v_i(k)] - v_i(k) \le \delta_i v_i(k)$, which implies that $e(k) = \Delta(k)K\hat{x}(k)$, and $\Delta(k) = \text{diag}\{\Delta_1(k), \Delta_2(k), \ldots, \Delta_m(k)\}$ with $\Delta_i(k) \in [-\delta_i, \delta_i]$.

Defining $\tilde{x}(k) = x(k) - \hat{x}(k)$, we can obtain the following closed-loop system:

$$\bar{x}(k + 1) = \tilde{A}\bar{x}(k), \tag{3.5}$$

where $\bar{x}(k) = [x^T(k), \tilde{x}^T(k)]^T$ and

$$\tilde{A} = \begin{bmatrix} A + B(I + \Delta(k))K & -B(I + \Delta(k))K \\ 0 & A - LC \end{bmatrix}.$$

It is well known that closed-loop system (3.5) is quadratically stable if and only if there exists a symmetric positive definite matrix $P > 0$ satisfying $\tilde{A}^T P \tilde{A} - P < 0$ [35]. However, in this matrix inequality, the feedback gain matrix K and observer gain matrix L are coupled, which leads to difficulty in determining K and L. The existing results, such as using two-step LMI approach [105], have conservatism to some degree. In order to circumvent this problem, we first give the following lemma.

Lemma 3.1. *The following statements are equivalent.*

(S1) There exist symmetric positive definite matrix $P > 0$, gain matrices K and L satisfying the following matrix inequality:

$$\begin{bmatrix} -P & \tilde{A}^T P \\ * & -P \end{bmatrix} < 0, \tag{3.6}$$

where $\tilde{A}$ is given in (3.5) and $*$ denotes the symmetric term in a symmetric matrix.

(S2) There exist symmetric positive definite matrices $P_{11} > 0$ and $Q_{22} > 0$, gain matrices K and L satisfying the following matrix inequalities:

$$\begin{bmatrix} -P_{11} & (A + B(I + \Delta(k))K)^T P_{11} \\ * & -P_{11} \end{bmatrix} < 0, \tag{3.7}$$

$$\begin{bmatrix} -Q_{22} & Q_{22}(A - LC)^T \\ * & -Q_{22} \end{bmatrix} < 0. \tag{3.8}$$

Proof. (S1)⇒(S2): Suppose there exists P such that (3.6) holds. Partitioning P as $P \triangleq \begin{bmatrix} P_{11} & P_{12} \\ P_{12}^T & P_{22} \end{bmatrix}$, and substituting it into (3.6) yields that

$$\begin{bmatrix} -P_{11} & -P_{12} & (A+B(I+\Delta(k))K)^T P_{11} \\ * & -P_{22} & -K^T(I+\Delta(k))B^T P_{11} + (A-LC)^T P_{12}^T \\ * & * & -P_{11} \\ * & * & * \end{bmatrix}$$

$$\left. \begin{matrix} (A+B(I+\Delta(k))K)^T P_{12} \\ -K^T(I+\Delta(k))B^T P_{12} + (A-LC)^T P_{22} \\ -P_{12} \\ -P_{22} \end{matrix} \right] < 0. \qquad (3.9)$$

Pre- and post-multiplying (3.9) by $\begin{bmatrix} I & 0 & 0 & 0 \\ 0 & 0 & I & 0 \end{bmatrix}$ and its transpose, respectively, yields (3.7) immediately.

Next, to show that (3.8) holds, we define $Q = P^{-1}$, and partition Q as $Q = \begin{bmatrix} Q_{11} & Q_{12} \\ Q_{12}^T & Q_{22} \end{bmatrix}$. Then, pre- and post-multiplying (3.6) by $\begin{bmatrix} Q & 0 \\ 0 & Q \end{bmatrix}$, we have

$$\begin{bmatrix} -Q & Q\tilde{A}^T \\ * & -Q \end{bmatrix} < 0. \qquad (3.10)$$

Substituting $\tilde{A}$ and Q into (3.10) yields that

$$\begin{bmatrix} -Q_{11} & -Q_{12} & Q_{11}(A+B(I+\Delta(k))K)^T - Q_{12}K^T(I+\Delta(k))B^T \\ * & -Q_{22} & Q_{12}^T(A+B(I+\Delta(k))K)^T - Q_{22}K^T(I+\Delta(k))B^T \\ * & * & -Q_{11} \\ * & * & * \end{bmatrix}$$

$$\left. \begin{matrix} Q_{12}(A-LC)^T \\ Q_{22}(A-LC)^T \\ -Q_{12} \\ -Q_{22} \end{matrix} \right] < 0. \qquad (3.11)$$

Pre- and post-multiplying (3.11) by $\begin{bmatrix} 0 & I & 0 & 0 \\ 0 & 0 & 0 & I \end{bmatrix}$ and its transpose, respectively, yields (3.8) immediately.

(S2)⇒(S1): First, we denote that

$$\Upsilon = \begin{bmatrix} -P_{11} & (A+B(I+\Delta(k))K)^T P_{11} \\ * & -P_{11} \end{bmatrix},$$

$$\Delta = \begin{bmatrix} -P_{22} & (A-LC)^T P_{22} \\ * & -P_{22} \end{bmatrix},$$

and it can be easily obtained that $\Upsilon < 0$ and $\Delta < 0$.

From Schur complement lemma, it can be verified that there exists a scalar $\epsilon_0 > 0$ such that the following inequality

$$\begin{bmatrix} \Upsilon & \Lambda^T \\ \Lambda & \epsilon\Delta \end{bmatrix} < 0, \tag{3.12}$$

holds for all $\epsilon \geq \epsilon_0$, where $\Lambda = \begin{bmatrix} 0 & 0 \\ -P_{11}B(I + \Delta(k))K & 0 \end{bmatrix}$. Exchanging of rows and columns, (3.12) yields exactly (3.6) with $P = \begin{bmatrix} P_{11} & 0 \\ 0 & \epsilon P_{22} \end{bmatrix}$. This completes the proof of the theorem. □

Remark 3.1. *It is easy to see that in Lemma 3.1, condition (S2) gives a single-step approach to analyze the asymptotic stability of closed-loop system (3.5), in which the matrices K and L are decoupled, which facilitates the design of observer-based the controller.*

Note that condition (3.7) contains of a matrix inequality defined over a rectangular parallelepiped, which is difficult to check. In [35], the authors regarded $\Delta(k)$ as norm-bounded uncertainties, and applied a well-known bounding inequality to eliminate $\Delta(k)$, which results in δ-independent conditions. In the following, we will present, instead, an equivalent expression of (3.7).

Lemma 3.2. *The following statements are equivalent.*

(S1) There exists a symmetric positive definite matrix $P_{11} > 0$ satisfying (3.7) for $\Delta(k) = \mathrm{diag}\{\Delta_1(k), \Delta_2(k), \ldots, \Delta_m(k)\}$ with $\forall \Delta_i(k) \in [-\delta_i, \delta_i]$.

(S2) There exists a symmetric positive definite matrix $P_{11} > 0$ satisfying the following LMIs:

$$\Omega_i \triangleq \begin{bmatrix} -P_{11} & (A + B(I + \Delta^{(i)})K)^T P_{11} \\ * & -P_{11} \end{bmatrix} < 0, \; i = 1, 2, \ldots, 2^m \tag{3.13}$$

where $\Delta^{(i)}$, $i = 1, 2, \ldots, 2^m$, represent the 2^m distinct matrices given in terms of $\mathrm{diag}\{\pm\delta_1, \pm\delta_2, \ldots, \pm\delta_m\}$.

Proof. First, observing that $\Delta_i(k) \in [-\delta_i, \delta_i]$, thus we can rewrite $\Delta(k)$ in (3.4) as the following polytopic form:

$$\Delta(k) = \sum_{i=1}^{2^m} \alpha_i(k)\Delta^{(i)}, \quad \sum_{i=1}^{2^m} \alpha_i(k) = 1, \quad \alpha_i(k) \geq 0. \tag{3.14}$$

(S1)⇒(S2): Taking $\Delta_i(k) = -\delta_i$ or δ_i in (3.7) for $i = 1, 2, \ldots, 2^m$, respectively, we obtain (3.13) immediately.
(S2)⇒(S1): Denote

$$\Omega \triangleq \begin{bmatrix} -P_{11} & (A + B(I + \Delta(k))K)^T P_{11} \\ * & -P_{11} \end{bmatrix}.$$

Noticing (3.14), it can be verified that $\Omega = \sum_{i=1}^{2^m} \alpha_i(k)\Omega_i < 0$. □

Remark 3.2. *When system (3.1) is with single input, that is, $m = 1$, the $\alpha_1(k)$ and $\alpha_2(k)$ can be chosen as $\alpha_1(k) = \frac{\delta + \Delta(k)}{2\delta}$ and $\alpha_2(k) = \frac{\delta - \Delta(k)}{2\delta}$.*

Remark 3.3. *It can be seen that by using a polytopic argument, the matrix inequality (3.7) defined on a polytope is equivalent to the matrix inequalities in (3.13) defined on the corresponding vertices. Nevertheless, the computational complexity may still be high if m is large.*

With the aid of Lemmas 3.1 and 3.2, we can derive the following theorem easily on designing the controller gain K and the observer gain L. The proof is straightforward and hence omitted.

Theorem 3.1. *Consider the discrete-time linear system (3.1). Given the logarithmic quantizer in (3.2) with quantization density ρ_i, $i = 1, 2, \ldots, m$, there exists an observer-based controller in the form of (3.3) such that the closed-loop system (3.5) is asymptotically stable if there exist matrices $\bar{P}_{11} > 0$, $\bar{Q}_{22} > 0$, $\bar{K}$ and $\bar{L}$ satisfying*

$$\begin{bmatrix} -\bar{P}_{11} & \bar{P}_{11}A^T + \bar{K}^T(I + \Delta^{(i)})B^T \\ * & -\bar{P}_{11} \end{bmatrix} < 0, \quad i = 1, 2, \ldots, 2^m \qquad (3.15)$$

$$\begin{bmatrix} -\bar{Q}_{22} & A^T\bar{Q}_{22} - C^T\bar{L}^T \\ * & -\bar{Q}_{22} \end{bmatrix} < 0, \qquad (3.16)$$

where $\Delta^{(i)}$, $i = 1, 2, \ldots, 2^m$, represent the 2^m distinct matrices given in terms of $\mathrm{diag}\{\pm\delta_1, \pm\delta_2, \ldots, \pm\delta_m\}$.

Furthermore, if the above conditions are feasible, the controller gain K and observer gain L are given by $K = \bar{K}\bar{P}_{11}^{-1}$, $L = \bar{Q}_{22}^{-1}\bar{L}$.

Remark 3.4. *It is worth pointing out that when the parameters δ_i, $i = 1, 2, \ldots, m$, are known, the conditions in Theorem 3.1 are linear matrix inequalities (LMIs) over the matrix variables $\bar{P}_{11} > 0$, $\bar{Q}_{22} > 0$, $\bar{K}$, and $\bar{L}$. However, when we employ Theorem 3.1 to compute the coarsest quantization density $\delta_i^{\max}$, which guarantees the closed-loop system to be quadratically stable, (3.15) and (3.16) are bilinear matrix inequalities (BMIs). To find $\delta_i^{\max}$, $i = 1, 2, \ldots, m$, we present one possible approach. First, we use line search, such as bisection method, to find $\delta_1^{\max}$ for given δ_i, $i = 2, \ldots, m$, and then fix $\delta_1^{\max}$, and search $\delta_2^{\max}$ for given δ_i, $i = 3, \ldots, m$. Repeating this procedure we can find $\delta_i^{\max}$, $i = 1, 2, \ldots, m$.*

3.4 REDUCED-ORDER OBSERVER-BASED QUANTIZED CONTROL

It is worth mentioning that, in many cases, at least one state variable can be either measured directly or calculated easily from the output. Then, if one or more state variables can be measured or observed directly, we can design a reduced-order observer that has an order lower than that of the plant to estimate the unmeasurable states, and a direct feedback path can be used to obtain the measured state values.

In this section, we will consider the case that partial information of the state vector will be obtained from output measurement $y(k)$, and quantized controller based on reduced-order observer will be designed. For simplicity, we assume that $C = [\ I \ \ 0\ ]$. Now, partition $x(k)$ as $x(k) = [x_1^T(k), x_2^T(k)]^T$, where $x_1(k) \in \mathbb{R}^q$ and $x_2(k) \in \mathbb{R}^{(n-q)}$, and accordingly, A can be partitioned as $A = \begin{bmatrix} A_{11} & A_{12} \\ A_{21} & A_{22} \end{bmatrix}$, where A_{11}, A_{12}, A_{21}, and A_{22} are with appropriate dimensions. Thus, system (3.1) can be rewritten as

$$x_1(k+1) = A_{11}x_1(k) + A_{12}x_2(k) + B_1u(k)$$
$$x_2(k+1) = A_{21}x_1(k) + A_{22}x_2(k) + B_2u(k)$$
$$y(k) = x_1(k)$$

By defining a new output $z(k) = y(k+1) - A_{11}x_1(k) - B_1u(k)$, we have $z(k) = A_{12}x_2(k)$, thus, the reduced-order observer can be chosen as

$$\hat{x}_2(k+1) = A_{22}\hat{x}_2(k) + \bar{u}(k) + L_r(z(k) - \hat{z}(k))$$
$$\hat{z}(k) = A_{12}\hat{x}_2(k)$$

where $\bar{u}(k) = A_{21}x_1(k) + B_2u(k)$, $L_r \in \mathbb{R}^{q \times (n-q)}$ is the observer gain.

The quantized controller can be designed as

$$u(k) = q[v(k)],$$
$$v(k) = K\xi(k), \tag{3.17}$$

where $q[\cdot]$ is the logarithmic quantizer as before, K is the controller gain to be determined later, and

$$\xi(k) = \begin{bmatrix} y(k) \\ \hat{x}_2(k) \end{bmatrix} = \begin{bmatrix} x_1(k) \\ \hat{x}_2(k) \end{bmatrix}.$$

Defining $\tilde{x}_2(k) = x_2(k) - \hat{x}_2(k)$, and considering the quantization error $e(k)$ in (3.4), we have the following closed-loop system:

$$\bar{x}_r(k+1) = \bar{A}_r\bar{x}_r(k) \tag{3.18}$$

where $\bar{x}_r(k) = [\xi^T(k), \tilde{x}_2^T(k)]^T$ and

$$\bar{A}_r = \begin{bmatrix} A + B(I + \Delta(k))K & \bar{L}A_{12} \\ 0 & A_{22} - L_rA_{12} \end{bmatrix}.$$

Next, similar to Theorem 3.1, we have the following theorem.

Theorem 3.2. *Consider the discrete-time linear system (3.1). Given the logarithmic quantizer in (3.2) with quantization density ρ_i, $i = 1, 2, \ldots, m$, there exists a reduced-observer-based controller in the form of (3.17) such that the closed-loop system (3.18) is asymptotically stable if there exist matrices $\tilde{P}_{11} > 0$, $\tilde{Q}_{22} > 0$, $\tilde{K}$ and $\tilde{L}$ satisfying*

$$\begin{bmatrix} -\tilde{P}_{11} & \tilde{P}_{11}A^T + \tilde{K}^T(I + \Delta^{(i)})B^T \\ * & -\tilde{P}_{11} \end{bmatrix} < 0, \ i = 1, 2, \ldots, 2^m \tag{3.19}$$

$$\begin{bmatrix} -\tilde{Q}_{22} & A_{22}^T\tilde{Q}_{22} - A_{12}^T\tilde{L}^T \\ * & -\tilde{Q}_{22} \end{bmatrix} < 0. \tag{3.20}$$

Furthermore, if the above conditions are feasible, the controller gain K and observer gain L_r are given by $K = \tilde{K}\tilde{P}_{11}^{-1}$, $L_r = \tilde{Q}_{22}^{-1}\tilde{L}$.

3.5 NUMERICAL SIMULATION

Example 3.1. *In this example, we consider a satellite system shown in Figure 3.3. The satellite system consists of two rigid bodies joined by a flexible link, which is*

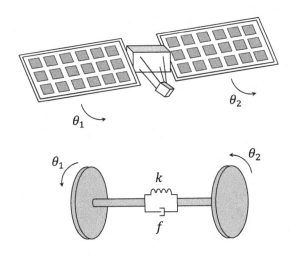

Figure 3.3 Satellite system

modeled as a spring with torque constant k and viscous damping f. Denoting the yaw angles for the two bodies (the main body and the instrumentation module) by θ_1 and θ_2, the control torque by $u(t)$, the moments of inertia of the two bodies by J_1 and J_2, the dynamic equations are given by

$$J_1\ddot{\theta}_1(t) + f(\dot{\theta}_1(t) - \dot{\theta}_2(t)) + k(\theta_1(t) - \theta_2(t)) = u(t),$$
$$J_2\ddot{\theta}_2(t) + f(\dot{\theta}_1(t) - \dot{\theta}_2(t)) + k(\theta_1(t) - \theta_2(t)) = 0.$$

A state space representation of the above equation is given by

$$diag\{1, 1, J_1, J_2\} \begin{bmatrix} \dot{\theta}_1(t) \\ \dot{\theta}_2(t) \\ \ddot{\theta}_1(t) \\ \ddot{\theta}_2(t) \end{bmatrix} = \begin{bmatrix} 0 & 0 & 1 & 0 \\ 0 & 0 & 0 & 1 \\ -k & k & -f & f \\ k & -k & f & -f \end{bmatrix} \begin{bmatrix} \theta_1(t) \\ \theta_2(t) \\ \dot{\theta}_1(t) \\ \dot{\theta}_2(t) \end{bmatrix} + \begin{bmatrix} 0 \\ 0 \\ u(t) \\ 0 \end{bmatrix}.$$

Here we choose $J_1 = J_2 = 1$ kg·m^2, $k = 0.09$ N·m, and $f = 0.04$ (the values of k and f are chosen within their respective ranges). It is assumed that only $\theta_2(t)$ can be measured online. Then the corresponding system matrices are given by

$$A_t = \begin{bmatrix} 0 & 0 & 1 & 0 \\ 0 & 0 & 0 & 1 \\ -0.09 & 0.09 & -0.04 & 0.04 \\ 0.09 & -0.09 & 0.04 & -0.04 \end{bmatrix}, \quad B_t = \begin{bmatrix} 0 \\ 0 \\ 1 \\ 0 \end{bmatrix}, \quad C_t = \begin{bmatrix} 0 & 1 & 0 & 0 \end{bmatrix}.$$

We can obtain the following discrete-time model with sampling period $T_s = 0.5s$,

$$A = \begin{bmatrix} 0.9889 & 0.0111 & 0.4932 & 0.0068 \\ 0.0111 & 0.9889 & 0.0068 & 0.4932 \\ -0.0438 & 0.0438 & 0.9695 & 0.0305 \\ 0.0438 & -0.0438 & 0.0305 & 0.9695 \end{bmatrix}, B = \begin{bmatrix} 0.1239 \\ 0.0011 \\ 0.4932 \\ 0.0068 \end{bmatrix}.$$

Our purpose is to design an observer-based quantized controller such that the closed-loop system is asymptotically stable. By using Theorem 3.1 with $\rho = 0.5$, we can obtain the controller and observer gains,

$$K = \begin{bmatrix} -1.0630 & -0.2361 & -2.1501 & -1.7285 \end{bmatrix}, L = \begin{bmatrix} 0.9478 \\ 1.2887 \\ 0.1203 \\ 0.5447 \end{bmatrix}.$$

The state trajectories of the closed-loop system with the obtained controller are shown in Figure 3.4(a), where $x(0) = [0.2, \ 0.3, \ -0.3, \ -0.2]^T$, and the control input is shown in Figure 3.4(b). It can be seen that the closed-loop system is asymptotically stable.

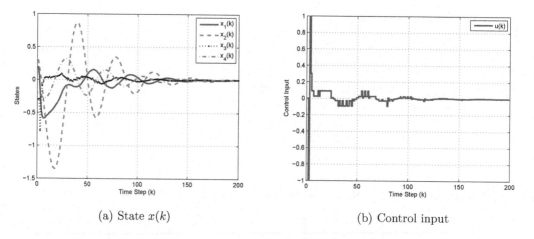

(a) State $x(k)$ (b) Control input

Figure 3.4 State and control input signals of satellite system (full-order case)

Furthermore, we take the state transformation $x_T(k) = Tx(k)$, where

$$T = \begin{bmatrix} 0 & 1 & 0 & 0 \\ 1 & 0 & 0 & 0 \\ 0 & 0 & 1 & 0 \\ 0 & 0 & 0 & 1 \end{bmatrix},$$

and system matrices A, B, and C become

$$A = \begin{bmatrix} 0.9889 & 0.0111 & 0.0068 & 0.4932 \\ 0.0111 & 0.9889 & 0.4932 & 0.0068 \\ 0.0438 & -0.0438 & 0.9695 & 0.0305 \\ -0.0438 & 0.0438 & 0.0305 & 0.9695 \end{bmatrix}, B = \begin{bmatrix} 0.0011 \\ 0.1239 \\ 0.4932 \\ 0.0068 \end{bmatrix}, C = \begin{bmatrix} 1 & 0 & 0 & 0 \end{bmatrix}.$$

By using Theorem 3.2, we can obtain the gain matrices K and L_r are

$$K = \begin{bmatrix} -0.2189 & -0.9624 & -2.0500 & -1.5579 \end{bmatrix}, \; L_r = \begin{bmatrix} 2.5618 \\ 0.4214 \\ 2.0465 \end{bmatrix},$$

and the simulation results are shown in Figure 3.5(a) and 3.5(b).

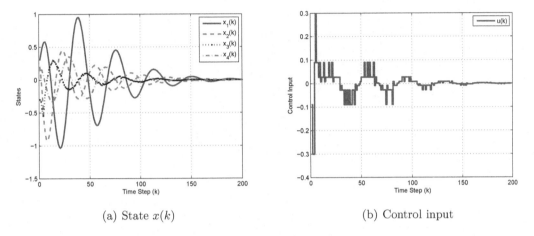

(a) State $x(k)$ (b) Control input

Figure 3.5 State and control input signals of satellite system (reduced-order case)

3.6 SUMMARY

The problem of output feedback control for discrete-time systems with quantized inputs has been investigated in this chapter. To design the observer and controller, the necessary and sufficient condition is established to guarantee the asymptotic stability of the closed-loop system. Furthermore, observer-based quantized feedback controller is also designed via a single-step approach. Finally, a numerical example is exploited to illustrate the applicability and effectiveness of the controller design methodologies proposed in this chapter.

State-Feedback Networked Delay Compensation Control for CPSs with Time-Varying Delay

4.1 INTRODUCTION

It is a well-known fact that, in most communication networks, data can be transmitted in "packet," which makes it possible in CPSs to actively compensate for the network communication delay and data dropout by sending a sequence of control input predictions (CIPs) in one data packet and then selecting the appropriate one corresponding to the current network condition. With this consideration, networked delay compensation control [also named networked predictive control (NPC)] strategies are recently proposed in [99]. The main feature is to predict the future control inputs of the system and take the corresponding control action according to the current network condition, and thus the network communication delay and data dropout can be compensated actively. Further studies on the networked delay compensation control methods are made in [100, 101, 228, 226, 225, 156, 154, 207, 204]. For example, in [226, 225], some networked delay compensation control approaches are proposed for networked Hammerstein-type systems. In [100, 101], the state space model based networked delay compensation control methods are developed for CPSs with feedback channel delay and both forward and feedback channel delays, respectively. In [126], the networked delay compensation control approach is developed for uncertain constrained nonlinear systems. In [77], an event-triggered networked delay compensation control method is presented for single input single output (SISO) CPSs.

This chapter addresses the problem of networked delay compensation control and stability analysis for CPSs with time varying network communication delay. As for networked delay compensation control scheme, two important problems should be solved, one is how to construct the CIPs, and the other is how to analyze the stability of the closed-loop networked control system. In this chapter, by taking the

DOI: 10.1201/9781003260882-4

full advantage of the packet-based transmission in CPSs, a state-based networked delay compensation control approach is proposed to actively compensate the network communication delay. Based on a switched system approach, stability analysis result is also established via average dwell time technique. Finally, the effectiveness of the proposed method is illustrated by a practical experiment.

4.2 SYSTEM DESCRIPTION

In this chapter, we consider the problem of networked delay compensation control for discrete-time control system. The remote controlled plant can be described as

$$x(k+1) = Ax(k) + Bu(k), \tag{4.1}$$

where $x(k) \in \mathbb{R}^n$ and $u(k) \in \mathbb{R}^m$ are the state vector and control input vector, respectively, and A and B are two constant matrices.

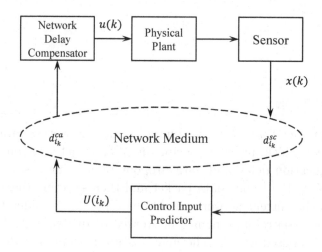

Figure 4.1 Block diagram of state-feedback networked delay compensation control

As shown in Figure 4.1, the following assumptions for system (4.1) are necessary throughout this chapter.

Assumption 4.1. *The sensor is time-triggered with sampling period T_s, and the controller and actuator are event-triggered.*

Assumption 4.2. *In all the data transmissions, each data packet is time-stamped. By comparing the time-stamp associated with the received data packets, the actuator always uses the most recent data packet and discards the old data packet, i.e., the actuator does not update the control input until a newer control packet arrives.*

Based on the above two assumptions, we call the received data packet is *effective*, if it is actually implemented to the plant after the comparison process. Denote the time stamp of the k-th effective data packet by i_k, and the corresponding round trip delay is d_{i_k}. As for d_{i_k}, we make the following assumption.

Assumption 4.3. *The round trip delay d_{i_k} is bounded, i.e., $d_m \leq d_{i_k} \leq d_M$, and d_m and d_M are constant positive scalars representing the lower and upper bounds, respectively, and $d_{i_k} = d_{i_k}^{sc} + d_{i_k}^{ca}$, where $d_{i_k}^{sc}$ and $d_{i_k}^{ca}$ are the sensor-to-controller channel delay and the controller-to-actuator channel delay, respectively.*

Remark 4.1. *In real applications, based on the commonly used network protocols, if a data packet does not arrive at a destination within a certain transmission time, then this data packet is considered lost. From a physical point of view, it is natural to assume that the network communication delay is bounded.*

Then, by considering the effect of the network transmission delay in CPSs, the real control system can be modeled as:

$$x(k+1) = Ax(k) + Bu(k)$$
$$u(k) = u(i_k),$$

where $u(i_k)$ denotes the control input at the k-th step.

The objective of this chapter is to propose the state feedback networked delay compensation control approach and stability analysis results for CPSs with time varying network communication delay.

4.3 DELAY COMPENSATION CONTROL SCHEME

In this section, we will focus on how to compensate the effects induced by network communication delay. Different from traditional control systems, in CPSs framework, the control loop is closed through the communication network, and we need to take into account both the plant and the communication channel. It is worth mentioning that the main feature of network is that a packet of data can be transmitted through the network at a time. Thus, we can utilize this important characteristic to design the networked controllers. The main idea is, at the controller side, the current control input and CIPs to be implemented in a finite number of future time instants are packetized into a data packet which is transmitted to the actuator side. At actuator side, one can select appropriate control signal to actively compensate for the current network communication delay.

Remark 4.2. *In an often used Ethernet IEEE 802.3 frame, the size of the effective load of the data packet is 368-bits, while for an 8-bits single step control signal, which can encode $2^8 = 256$ different control actions being ample for most control implementations, in this case, 45 steps of controller-to-actuator channel delay is allowed, which can actually meet the requirements of most practical control systems. Therefore, the packet-based data transmission in CPSs enables us to transmit a sequence of control signals simultaneously.*

Now, we suppose that the state information sampled at $i_k T_s$ is successfully transmitted to controller. At the controller side, therefore, we can choose the CIPs set as

$$U(i_k) = \begin{bmatrix} u(i_k) & u(i_k+1) & u(i_k+2) & \cdots & u(i_k+d_M) \end{bmatrix}. \tag{4.2}$$

Once $U(i_k)$ is received at the actuator side, we can apply the $(d_{i_k} + 1)$-th control input value in CIP $U(i_k)$ to the plant, where d_{i_k} is the round trip delay associated with $U(i_k)$. Then, the closed-loop system can be rewritten as

$$x(k+1) = Ax(k) + Bu(k)$$
$$u(k) = u(i_k + d_{i_k}),$$

One important problem associated with networked delay compensation control approach is how to determine the CIPs in (4.2). Next, we assume the state information can be obtained directly, and present one possible way to construct the CIPs.

Step 1: The current control input can be chosen as state feedback as follows

$$u(i_k) = Kx(i_k),$$

where K can be designed by modern control method, for example, linear quadratic regulator (LQR), pole assignment, etc, guaranteeing $A + BK$ is Schur stable;

Step 2: Based on $u(i_k)$ and system model of (4.1), we have

$$\hat{x}(i_k + 1) = Ax(i_k) + Bu(i_k)$$
$$= (A + BK)x(i_k)$$
$$u(i_k + 1) = K\hat{x}(i_k + 1),$$

where $\hat{x}$ is an auxiliary state variable;

Step i: By some iterations, we have

$$\hat{x}(i_k + i - 1) = A\hat{x}(i_k + i - 2) + Bu(i_k + i - 2)$$
$$= (A + BK)\hat{x}(i_k + i - 2)$$
$$= (A + BK)^{i-1}x(i_k)$$
$$u(i_k + i - 1) = K\hat{x}(i_k + i - 1),$$

Step $d_M + 1$: Finally, with $u(i_k + d_M - 1)$, we have

$$\hat{x}(i_k + d_M) = A\hat{x}(i_k + d_M - 1) + Bu(i_k + d_M - 1)$$
$$= (A + BK)\hat{x}(i_k + d_M - 1)$$
$$= (A + BK)^{d_M}x(i_k)$$
$$u(i_k + d_M) = K\hat{x}(i_k + d_M).$$

Thus, we can obtain the following CIP

$$U(i_k) = \begin{bmatrix} Kx(i_k) & K(A+BK)x(i_k) & \cdots & K(A+BK)^{d_M}x(i_k) \end{bmatrix}.$$

After finishing the calculations, the CIP can now be transmitted to actuator side in a single data packet by the network.

Remark 4.3. *It is worth noticing that, different from the procedure established in [155], in the construction of CIPs, the auxiliary variables are introduced since there are no new states of the plant is received except $x(i_k)$ at the current time step.*

4.4 STABILITY ANALYSIS

If $U(i_k)$ is effective, we will select the control input as $K(A+BK)^{d_{i_k}}x(i_k)$ and apply it to the plant. Then, the closed-loop system can be written as

$$x(k+1) = Ax(k) + BK(A+BK)^{d_{i_k}}x(i_k), \tag{4.3}$$

where $d_{i_k} \in \mathcal{I} \triangleq \{d_m, d_m+1, \dots, d_M\}$.

Let $\tau_k = k - i_k$, $k = 1, 2, 3, \dots$, and assume that $\tau_m \leq \tau_k \leq \tau_M$, where τ_m and τ_M are two constant scalars. Then system (4.3) can be further rewritten as the following time delay system:

$$x(k+1) = Ax(k) + BK(A+BK)^{d_{i_k}}x(k-\tau_k). \tag{4.4}$$

It can be seen that, when d_{i_k} takes value in $\mathcal{I}$, system (4.4) is a switched time delay system. Let $\sigma_k : [0, +\infty) \to \mathcal{I}$ be the switching signal. Therefore, we rewrite system (4.4) as switched time delay system, and the i-th subsystem can be given as follow:

$$x(k+1) = Ax(k) + A_i x(k-\tau_k), \ i \in \mathcal{I}, \tag{4.5}$$

where $A_i = BK(A+BK)^i$.

For simplicity, we give the following notations:

$$\eta(k) = x(k+1) - x(k), \ \xi(k) = \begin{bmatrix} x^T(k) & x^T(k-\tau_k) \end{bmatrix}^T,$$

$$\zeta(k) = \begin{bmatrix} \xi^T(k) & x^T(k-\tau_m) & x^T(k-\tau_M) \end{bmatrix}^T,$$

$$\Xi_{1i} = \begin{bmatrix} A & A_i & 0 & 0 \end{bmatrix}, \ \Xi_{2i} = \begin{bmatrix} A-I & A_i & 0 & 0 \end{bmatrix}. \tag{4.6}$$

Thus, we have $x(k+1) = \Xi_{1i}\zeta(k)$ and $\eta(k) = \Xi_{2i}\zeta(k)$.

Here, we choose the following Lyapunov function candidate for the i-th subsystem of switched system (4.5),

$$V_i(k) = V_{1i}(k) + V_{2i}(k) + V_{3i}(k) + V_{4i}(k),$$

where

$$V_{1i}(k) = x^T(k)P_i x(k)$$

$$V_{2i}(k) = \sum_{s=k-\tau_m}^{k-1} x^T(s)\lambda^{\alpha(s-k+1)}Q_{1,i}x(s) + \sum_{s=k-\tau_M}^{k-1} x^T(s)\lambda^{\alpha(s-k+1)}Q_{2,i}x(s)$$

$$V_{3i}(k) = \sum_{\theta=-\tau_M+1}^{0} \sum_{s=k-1+\theta}^{k-1} \eta^T(s)\lambda^{\alpha(s-k+1)}R_{1,i}\eta(s)$$

$$+ \sum_{\theta=-\tau_M+1}^{-\tau_m} \sum_{s=k-1+\theta}^{k-1} \eta^T(s)\lambda^{\alpha(s-k+1)}R_{2,i}\eta(s)$$

$$V_{4i}(k) = \sum_{\theta=-\tau_M+1}^{-\tau_m+1} \sum_{s=k-1+\theta}^{k-1} x^T(s)\lambda^{\alpha(s-k+1)}Q_{3,i}x(s).$$

The following definition and lemma will be used in the derivation of the main results.

Definition 4.1. *[215] The equilibrium $x^* = 0$ of system (4.5) is said to be exponentially stable if the solution $x(k)$ of system (4.5) satisfies*

$$\|x(k)\| \leq \kappa \lambda^{-(k-k_0)} \|x(k_0)\|_l, \ \forall k \geq k_0$$

for any initial condition $x(k_0)$, where $\|\cdot\|$ denotes the Euclidean norm, and $\|x(k_0)\|_l = \sup_{k_0 - \tau_M \leq \theta \leq k_0} \{x(\theta)\}$, κ is the decay coefficient, $\lambda > 1$ is the decay rate.

Based on the time delay system theory [66], we have the following lemma.

Lemma 4.1. *Consider switched time delay system (4.5). For given constants $\alpha > 0$ and $\lambda > 1$, if there exist matrices $P_i > 0$, $Q_{j,i} \geq 0$, $j = 1, 2, 3$, $R_{l,i} \geq 0$, $l = 1, 2$,*

$$X_i = \begin{bmatrix} X_{11i} & X_{12i} \\ * & X_{22i} \end{bmatrix} \geq 0, \ Y_i = \begin{bmatrix} Y_{11i} & Y_{12i} \\ * & Y_{22i} \end{bmatrix} \geq 0 \ \text{and matrices} \ M_i = \begin{bmatrix} M_{1i} \\ M_{2i} \end{bmatrix},$$

$$N_i = \begin{bmatrix} N_{1i} \\ N_{2i} \end{bmatrix} \ \text{and} \ S_i = \begin{bmatrix} S_{1i} \\ S_{2i} \end{bmatrix} \ \text{with appropriate dimensions, } i \in \mathcal{I}, \ \text{such that}$$

$$\begin{bmatrix} \Phi_i & \Xi_{1i}^T P_i & \sqrt{\tau_M} \Xi_{2i}^T R_{1,i}^T & \sqrt{\tau_M - \tau_m} \Xi_{2i}^T R_{2,i}^T \\ * & -P_i & 0 & 0 \\ * & * & -R_{1,i} & 0 \\ * & * & * & -R_{2,i} \end{bmatrix} < 0, \tag{4.7}$$

$$\Psi_{1i} = \begin{bmatrix} X_i & M_i \\ * & \lambda^{-\alpha \tau_M} R_{1,i} \end{bmatrix} \geq 0, \ \ \Psi_{2i} = \begin{bmatrix} Y_i & S_i \\ * & \lambda^{-\alpha(\tau_M + 1)} R_{2,i} \end{bmatrix} \geq 0, \tag{4.8}$$

$$\Psi_{3i} = \begin{bmatrix} X_i + Y_i & N_i \\ * & \lambda^{-\alpha(\tau_M + 1)} (R_{1,i} + R_{2,i}) \end{bmatrix} \geq 0, \tag{4.9}$$

where Ξ_{1i} and Ξ_{2i} are given in (4.6), and

$$\Phi_i = \begin{bmatrix} \Phi_{11i} & \Phi_{12i} & S_{1i} & -N_{1i} \\ * & \Phi_{22i} & S_{2i} & -N_{2i} \\ * & * & -\lambda^{-\alpha \tau_m} Q_{1,i} & 0 \\ * & * & * & -\lambda^{-\alpha \tau_M} Q_{2,i} \end{bmatrix}$$

$$\Phi_{11i} = Q_{1,i} + Q_{2,i} + (\tau_M - \tau_m + 1) Q_{3,i} - \lambda^{-\alpha} P_i$$
$$+ \tau_M X_{11i} + (\tau_M - \tau_m) Y_{11i} + M_{1i} + M_{1i}^T$$

$$\Phi_{12i} = d_M X_{12i} + (\tau_M - \tau_m) Y_{12i} - M_{1i} + N_{1i} + M_{2i}^T - S_{1i}$$

$$\Phi_{22i} = -\lambda^{-\alpha \tau_M} Q_{3,i} + d_M X_{22i} + (\tau_M - \tau_m) Y_{22i}$$
$$- M_{2i} - M_{2i}^T + N_{2i} + N_{2i}^T - S_{2i} - S_{2i}^T.$$

Then, along the trajectory of system (4.5), we have

$$V_i(k+1) \leq \lambda^{-\alpha} V_i(k). \tag{4.10}$$

Proof. Defining $\Delta V_i(k) = V_i(k+1) - \lambda^{-\alpha}V_i(k)$, then we have

$$\begin{aligned}
\Delta V_{1i}(k) &= x^T(k+1)Px(k+1) - \lambda^{-\alpha}x^T(k)Px(k) \\
&= \zeta^T(k)\Xi_1^T P\Xi_1\zeta(k) - \lambda^{-\alpha}x^T(k)Px(k),
\end{aligned}$$

$$\begin{aligned}
\Delta V_{2i}(k) &= x^T(k)Q_1 x(k) - x^T(k-d_m)\lambda^{-\alpha d_m}Q_1 x(k-d_m) \\
&+ x^T(k)Q_2 x(k) - x^T(k-d_M)\lambda^{-\alpha d_M}Q_2 x(k-d_M),
\end{aligned}$$

$$\begin{aligned}
\Delta V_{3i}(k) &= \zeta^T(k)\Xi_2^T(d_M R_1 + (d_M - d_m)R_2)\Xi_2\zeta(k) - \sum_{s=k-d_k}^{k-1} \eta^T(s)\lambda^{\alpha(s-k)}R_1\eta(s) \\
&- \sum_{s=k-d_k}^{k-d_m-1} \eta^T(s)\lambda^{\alpha(s-k)}R_2\eta(s) - \sum_{s=k-d_M}^{k-d_k-1} \eta^T(s)\lambda^{\alpha(s-k)}(R_1 + R_2)\eta(s) \\
&\leq \zeta^T(k)\Xi_2^T(d_M R_1 + (d_M - d_m)R_2)\Xi_2\zeta(k) - \sum_{s=k-d_k}^{k-1} \eta^T(s)\lambda^{-\alpha d_M}R_1\eta(s) \\
&- \sum_{s=k-d_k}^{k-d_m-1} \eta^T(s)\lambda^{-\alpha d_M}R_2\eta(s) - \sum_{s=k-d_M}^{k-d_k-1} \eta^T(s)\lambda^{-\alpha d_M}(R_1 + R_2)\eta(s),
\end{aligned}$$

and

$$\begin{aligned}
\Delta V_{4i}(k) &= (d_M - d_m + 1)x^T(k)Q_3 x(k) - \sum_{s=k-d_M}^{k-d_m} x^T(s)\lambda^{\alpha(s-k)}Q_3 x(s) \\
&\leq (d_M - d_m + 1)x^T(k)Q_3 x(k) - x^T(k-d_k)\lambda^{-\alpha d_M}Q_3 x(k-d_k),
\end{aligned}$$

and note that, for any matrices M_i, N_i, S_i with appropriate dimensions, the following three equations hold,

$$2\xi^T(k)M_i\left[x(k) - x(k-\tau_k) - \sum_{s=k-\tau_k}^{k-1} \eta(s)\right] = 0,$$

$$2\xi^T(k)N_i\left[x(k-\tau_k) - x(k-\tau_M) - \sum_{s=k-\tau_M}^{k-\tau_k-1} \eta(s)\right] = 0,$$

$$2\xi^T(k)S_i\left[x(k-\tau_m) - x(k-\tau_k) - \sum_{s=k-\tau_k}^{k-\tau_m-1} \eta(s)\right] = 0,$$

and, for any matrices X_i and Y_i with appropriate dimensions, the following equations are true:

$$0 = \sum_{s=k-\tau_M}^{k-1} \xi^T(k)X_i\xi(k) - \sum_{s=k-\tau_M}^{k-1} \xi^T(k)X\xi(k)$$

$$= \tau_M\xi^T(k)X_i\xi(k) - \sum_{s=k-\tau_k}^{k-1} \xi^T(k)X_i\xi(k) - \sum_{s=k-\tau_M}^{k-\tau_k-1} \xi^T(k)X_i\xi(k),$$

$$0 = \sum_{s=k-d_M}^{k-\tau_m-1} \xi^T(k)Y\xi(k) - \sum_{s=k-\tau_M}^{k-\tau_m-1} \xi^T(k)Y\xi(k)$$

$$= (\tau_M - \tau_m)\xi^T(k)Y_i\xi(k) - \sum_{s=k-\tau_k}^{k-\tau_m-1} \xi^T(k)Y_i\xi(k) - \sum_{s=k-\tau_M}^{k-\tau_k-1} \xi^T(k)Y_i\xi(k).$$

Therefore, we have

$$\Delta V_i(k) = \zeta^T(k)[\Phi_i + \Xi_{1i}^T P_i \Xi_{1i} + \Xi_{2i}^T[\tau_M R_{1,i} + (\tau_M - \tau_m)R_{2,i}]\Xi_{2i}\zeta(k)$$
$$- \sum_{s=k-\tau_k}^{k-1} \delta_{k,s}^T \Psi_{1i}\delta_{k,s} - \sum_{s=k-\tau_k}^{k-\tau_m-1} \delta_{k,s}^T \Psi_{2i}\delta_{k,s} - \sum_{s=k-\tau_M}^{k-\tau_k-1} \delta_{k,s}^T \Psi_{3i}\delta_{k,s},$$

where $\delta_{k,s} = [\ \xi^T(k)\ \ \eta^T(s)\ ]^T$.

It follows from (4.7)–(4.9) and Schur complement lemma that $\Delta V_i(k) \leq 0$, which implies that $V_i(k+1) - \lambda^{-\alpha}V_i(k) \leq 0$. □

Definition 4.2. *[96] For any $T_2 > T_1 \geq 0$, let $N_\sigma(T_1, T_2)$ denote the number of switching of $\sigma(t)$ over (T_1, T_2). If $N_\sigma(T_1, T_2) \leq N_0 + (T_2 - T_1)/T_a$ holds for $T_a > 0$, $N_0 \geq 0$, then T_a is called average dwell time.*

Motivated by the stability analysis approach of switched systems [215], we give the stability analysis results of control system (4.4).

Theorem 4.1. *Consider networked closed-loop system (4.4). For given constants $\alpha > 0$, $\lambda > 1$, and $\mu > 1$, if there exist matrices $P_i > 0$, $Q_{j,i} \geq 0$, $j = 1, 2, 3$, $R_{l,i} \geq 0$, $l = 1, 2$, $X_i \geq 0$, $Y_i \geq 0$ and matrices M_i, N_i, and S_i with appropriate dimensions, $i \in \mathcal{I}$, satisfying (4.7)–(4.9) and*

$$P_i \leq \mu P_j,\ Q_{l,i} \leq \mu Q_{l,j}, l = 1, 2, 3,\ R_{s,i} \leq \mu R_{s,j}, s = 1, 2,\ \forall i, j \in \mathcal{I} \qquad (4.11)$$

hold, then networked closed-loop system (4.4) is exponentially stable for any switching signal with average dwell time T_a satisfying

$$T_a > T_a^* = \frac{\ln \mu}{\alpha \ln \lambda},$$

and the decay rate is $\sqrt{\lambda^\rho}$, where $\rho = -\frac{\ln \mu}{\alpha T_a \ln \lambda} + 1$.

Proof. Take the following Lyapunov function [66]:

$$V_{\sigma_k}(k) = V_{1,\sigma_k}(k) + V_{2,\sigma_k}(k) + V_{3,\sigma_k}(k) + V_{4,\sigma_k}(k),$$

where

$$V_{1,\sigma_k}(k) = x^T(k)P_{\sigma_k}x(k)$$

$$V_{2,\sigma_k}(k) = \sum_{s=k-\tau_m}^{k-1} x^T(s)\lambda^{\alpha(s-k+1)}Q_{1,\sigma_k}x(s) + \sum_{s=k-\tau_M}^{k-1} x^T(s)\lambda^{\alpha(s-k+1)}Q_{2,\sigma_k}x(s)$$

$$V_{3,\sigma_k}(k) = \sum_{\theta=-\tau_M+1}^{0} \sum_{s=k-1+\theta}^{k-1} \eta^T(s)\lambda^{\alpha(s-k+1)}R_{1,\sigma_k}\eta(s)$$

$$+ \sum_{\theta=-\tau_M+1}^{-\tau_m} \sum_{s=k-1+\theta}^{k-1} \eta^T(s)\lambda^{\alpha(s-k+1)}R_{2,\sigma_k}\eta(s)$$

$$V_{4,\sigma_k}(k) = \sum_{\theta=-\tau_M+1}^{-\tau_m+1} \sum_{s=k-1+\theta}^{k-1} x^T(s)\lambda^{\alpha(s-k+1)}Q_{3,\sigma_k}x(s)$$

and P_{σ_k}, Q_{1,σ_k}, Q_{2,σ_k}, Q_{3,σ_k}, R_{1,σ_k}, R_{2,σ_k} are the solutions of matrix inequalities (4.7)–(4.9). From (4.11), we have

$$V_{\sigma_{k_t}}(k_t) \leq \mu V_{\sigma_{k_t-1}}(k_t),$$

where k_t denotes the switching time instant. Noticing that $\sigma_{k_t-1} = \sigma_{k_{t-1}}$, we obtain

$$V_{\sigma_k}(k) \leq \lambda^{-\alpha(k-k_t)}V_{\sigma_{k_t}}(k_t)$$

$$\leq \lambda^{-\alpha(k-k_t)}\mu V_{\sigma_{k_t-1}}(k_t)$$

$$\leq \mu\lambda^{-\alpha(k-k_t)}\lambda^{-\alpha(k_t-k_{t-1})}V_{\sigma_{k_{t-1}}}(k_{t-1})$$

$$\vdots$$

$$\leq \mu^{N_\sigma[k,k_0]}\lambda^{-\alpha(k-k_t)}\lambda^{-\alpha(k_t-k_{t-1})}\cdots\lambda^{-\alpha(k_1-k_0)}V_{\sigma_{k_0}}(k_0)$$

$$= \mu^{N_\sigma[k,k_0]}\lambda^{-\alpha(k-k_0)}V_{\sigma_{k_0}}(k_0)$$

$$= \lambda^{N_\sigma[k,k_0]\frac{\ln\mu}{\ln\lambda}}\lambda^{-\alpha(k-k_0)}V_{\sigma_{k_0}}(k_0)$$

$$= \lambda^{-\alpha(k-k_0)\frac{N_\sigma[k,k_0]}{-\alpha(k-k_0)}\frac{\ln\mu}{\ln\lambda}}\lambda^{-\alpha(k-k_0)}V_{\sigma_{k_0}}(k_0)$$

$$\leq \lambda^{-\alpha(k-k_0)(-\frac{\ln\mu}{\alpha T_a \ln\lambda}+1)}V_{\sigma_{k_0}}(k_0)$$

$$= (\lambda^\rho)^{-\alpha(k-k_0)}V_{\sigma_{k_0}}(k_0).$$

Furthermore, there exist constants $a > 0$ and $b > 0$ satisfying the following inequality:

$$a\|x(k)\|^2 \leq V_{\sigma_k}(k) \leq (\lambda^\rho)^{-\alpha(k-k_0)}V_{\sigma_{k_0}}(k_0)$$

$$\leq (\lambda^\rho)^{-\alpha(k-k_0)}b\|x(k_0)\|_l^2.$$

Then, we have

$$\|x(k)\| \leq \sqrt{\frac{b}{a}}\sqrt{\lambda^\rho}^{-\alpha(k-k_0)}\|x(k_0)\|_l,$$

which implies system (4.4) is exponentially stable, and decay rate is $\sqrt{\lambda^\rho}$. □

Remark 4.4. *Theorem 4.1 implies that the exponential stability of the closed-loop system (4.4) is guaranteed if all its subsystems are exponentially stable and the average dwell time is large enough, at least is* $\frac{\ln \mu}{\alpha \ln \lambda}$.

Remark 4.5. *According to the proposed networked delay compensation control scheme, the control input of the plant is selected in the CIPs based on the round trip delay* d_{i_k}, *and when* d_{i_k} *taking value in* $\mathcal{I} = \{d_m, d_m + 1, \ldots, d_M\}$, *the closed-loop system can be modeled as a switched system (4.5). It should be pointed out that, when we construct the CIPs, it is naturally to assume that, no matter which element in the CIPs is implemented to the plant, the closed-loop system should be stable, which results in the situation that all the individual subsystems of switched system (4.5) are stable.*

Remark 4.6. *In this chapter, to construct the CIPs, the nonlinear terms* $BK(A + BK)^{d_{i_k}}$, $d_{i_k} \in \{d_m, d_m + 1, \ldots, d_M\}$, *of controller gain* K *are introduced, which leads to the difficulty in controller design. It is clear that, we cannot design the controller gain* K *via LMI techniques. Besides that, the existing efficient iterative algorithms, such as CCL algorithm, ILMI algorithm, cannot be used in this situation. In this case, we choose a relatively conservative approach to design a networked delay compensation controller, we first select a local controller gain* K *for the plant guaranteeing the* $A + BK$ *is Schur stable, and has certain desired properties, e.g., having a small damping ratio. Then, based on the selected controller gain* K, *the CIPs can be constructed via the proposed approach, and the established networked delay compensation control approach can be used in network environment. It is worth mentioning that, although no controller design is given in this chapter, more freedom on selecting controller gain* K *is guaranteed, which is more preferred for engineering application of CPSs.*

Remark 4.7. *In this chapter, the state feedback based networked delay compensation controller is proposed, and the stability is analyzed by using switched systems approach. However, when the states of the system are not measurable, we can consider the output feedback based networked delay compensation control. It should be pointed out that, if we choose the static output feedback* $u(k) = Ly(k)$, *the design approach in this chapter is also effective if we reconstruct* $U(i_k)$ *in (4.2) as*

$$U(i_k) = \begin{bmatrix} LCx(i_k) & LC(A + BLC)x(i_k) & \cdots & LC(A + BLC)^{d_M}x(i_k) \end{bmatrix},$$

where L *is chosen such that* $A + BLC$ *is Schur stable, and the stability analysis can be performed in a similar manner.*

4.5 PRACTICAL EXPERIMENT

In this section, we consider the problem of practical experiment. The plant is selected as the ball and beam system produced by Googol Technology Limited, as shown in Figure 4.2.

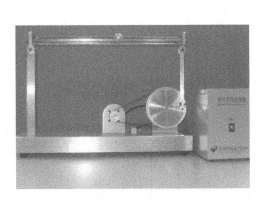 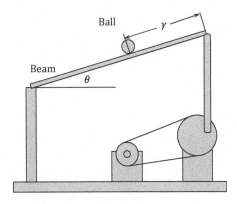

Figure 4.2 Ball and beam system and its schematic diagram

The nonlinear model of ball and beam system is

$$0 = \left(\frac{J_b}{R^2} + m\right)\ddot{\gamma} + mg\sin\theta - m\gamma\dot{\theta}^2,$$

$$\tau = (m\gamma^2 + J + J_b)\ddot{\theta} + 2m\gamma\dot{\gamma}\dot{\theta} + mg\gamma\cos\theta,$$

where τ is the torque applied to the beam. The parameter and physical meaning of beam and ball system are given in Table 4.1.

Table 4.1 The parameters of beam and ball system

Symbol	Meaning
J	Moment of inertia of beam
J_b	Moment of inertia of ball
m	Mass of ball
R	Radius of ball
L	Length of the beam
g	Gravitational constant

Define $b = \frac{m}{\frac{J_b}{R^2}+m}$, and change the coordinates in the input space using the invertible nonlinear transformation [60]:

$$\tau = 2m\gamma\dot{\gamma}\dot{\theta} + mg\gamma\cos\theta + (m\gamma^2 + J + J_b)u,$$

where $u = \ddot{\theta}$ is the control input, and then the system can be written in state space form as

$$\begin{bmatrix} \dot{x}_1 \\ \dot{x}_2 \\ \dot{x}_3 \\ \dot{x}_4 \end{bmatrix} = \begin{bmatrix} x_2 \\ b(x_1 x_4^2 - g\sin x_3) \\ x_4 \\ 0 \end{bmatrix} + \begin{bmatrix} 0 \\ 0 \\ 0 \\ 1 \end{bmatrix} u,$$

where $x = [x_1, x_2, x_3, x_4]^T = [\gamma, \dot{\gamma}, \theta, \dot{\theta}]^T$. The system parameters are $M = 0.11$ kg, $R = 0.015$ m, $J_b = 9.9 \times 10^{-6}$ kgm^2, $g = 9.81$ m/s^2, thus $b = 0.7143$.

Note that, at the equilibrium point $\theta = 0$, we have $mg \sin \theta - m\gamma\dot{\theta}^2 \approx mg\theta$, which means that $\dot{x}_2 = b(x_1 x_4^2 - g \sin x_3) \approx -bgx_3$. Thus, the linearized equation about the beam angle gives us the following linear approximation of the system:

$$\dot{x}(t) = A_t x(t) + B_t u(t),$$

where

$$A_t = \begin{bmatrix} 0 & 1 & 0 & 0 \\ 0 & 0 & -bg & 0 \\ 0 & 0 & 0 & 1 \\ 0 & 0 & 0 & 0 \end{bmatrix}, \ B_t = \begin{bmatrix} 0 \\ 0 \\ 0 \\ 1 \end{bmatrix}.$$

By taking the sampling period as $T_s = 0.02s$, we get the discretized system

$$x(k+1) = Ax(k) + Bu(k), \tag{4.12}$$

where

$$A = \begin{bmatrix} 1.0000 & 0.0200 & -0.0014 & -0.0000 \\ 0 & 1.0000 & -0.1401 & -0.0014 \\ 0 & 0 & 1.0000 & 0.0200 \\ 0 & 0 & 0 & 1.0000 \end{bmatrix}, B = \begin{bmatrix} -0.0000 \\ -0.0000 \\ 0.0002 \\ 0.0200 \end{bmatrix}.$$

Next, to verify the proposed networked delay compensation control scheme, we will perform the networked experiment for discrete system (4.12). To implement CPSs, a test rig is built, which includes two computers and one beam and ball system as the controlled plant, and its diagram is shown in Figure 4.3. The signals are physically transmitted between two intranet IP addresses 10.108.22.221 and 10.108.22.198 that are both located on the campus network, and TCP protocol is used. To simulate the Internet network communication delay, we add delay artificially into Intranet environment.

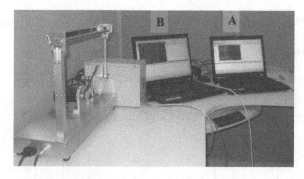

Figure 4.3 Networked control test rig

To show the advantages of proposed control scheme, we compare it with Proportional-Integral-Derivative (PID) control scheme. First, we choose the PID parameters via local experiment, $K_p = 8$, $K_i = 1$, and $K_d = 10$. Furthermore, we use LQR control to get the controller gain K, take $Q = \text{diag}\{1000, 0, 100, 0\}$, $R = 1$, then we can get $K = [28.5270, 20.0386, -49.3171, -9.7944]$ by Matlab toolbox. In this case, by using Theorem 4.1 and choosing $\lambda = 1.01$, $\alpha = 0.54$, $\mu = 1.01$, then, it can be verified the networked closed-loop system (4.4) is exponentially stable under the delay $1 \leq d_k \leq 3$, and average dwell time $T_a^* = 1.8519$.

Now, we examine the stability of the beam and ball system by using real experiment, the control objective is to stabilize the ball at the middle of the beam, i.e., $\gamma = 0.2$ m. First, we add 1–3 step delay into Intranet environment under the average dwell time $T_a = 3.33$ (case 1), the experiment result by using PID control scheme is shown in Figure 4.4(a). When utilizing the networked delay compensation control

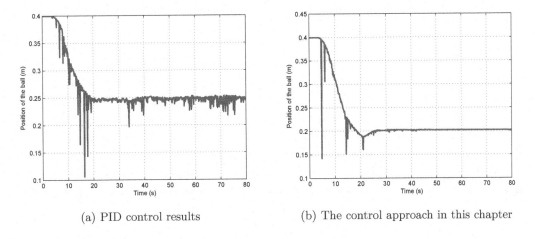

(a) PID control results   (b) The control approach in this chapter

Figure 4.4 Comparative experiment results for the position of the ball (case 1)

scheme proposed in this chapter, the experiment result is shown in Figure 4.4(b), from which we can see clearly that our control approach is better than PID control scheme.

Furthermore, we add 1–3 step delay into Intranet environment under the average dwell time $T_a = 2.5$ (Case 2), the experiment result is shown in Figure 4.5(a) for PID control scheme, we can find that it is difficult to stabilize the ball at one position of the beam. However, by using our proposed control approach, the experiment result is shown in Figure 4.5(b), we can find that our approach can also work well.

From practical experiments, it can be concluded that the proposed networked control scheme can compensate for the network communication delay of simulated Internet-based control systems effectively.

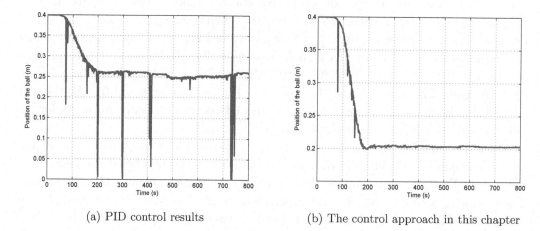

(a) PID control results   (b) The control approach in this chapter

Figure 4.5   Comparative experiment results for the position of the ball (case 2)

4.6   SUMMARY

The problems of networked delay compensation control and stability analysis have been investigated for CPSs. The network communication delay is assumed as time varying and bounded. First, a state-based networked delay compensation control approach is proposed to actively compensate the network communication delay. Then, by applying switched system approach and average dwell time technique, the stability analysis result is also presented. Finally, a practical experiment is included to demonstrate the effectiveness and potential of the theoretic results obtained.

CHAPTER 5

Output-Feedback Networked Delay Compensation Control for CPSs with Random Delay

5.1 INTRODUCTION

In Chapter 4, the state feedback networked delay compensation control strategies are proposed for CPSs with time-varying network communication delays. Nevertheless, in some practical applications, it is difficult or prohibitively expensive to measure the full state information, and only the output information can be physically measured. In this situation, it is necessary to design the output feedback controllers. To get around this issue, the state observers are often applied to estimate the system states, and then the observer-based control schemes can be established even if it possibly increases the complexity of the closed-loop systems. In [102, 103], the quantized observer-based output feedback control problems for CPSs are considered. In [101, 55], the observer-based output feedback delay compensation control problems are investigated for CPSs. Besides, the static output feedback control schemes using only the output information is an alternative approach. In [21, 212], the output feedback controllers are designed for CPSs with data dropout. In [137, 45], the dynamic output feedback controllers are designed for CPSs with network communication delays. Compared with the observer-based control approaches, the static output feedback control approaches are simpler and more economical. Moreover, it is well-known that, the Markov chain, a discrete-time stochastic process with the Markov property, can be effectively used to model network communication delays and data dropouts in CPSs [209, 137, 165, 175]. In [114], the network communication delays are modeled by using the Markov chains, and the linear quadratic Gaussian (LQG) optimal controller design method is proposed. In [174], both the state- and output-feedback networked controller design methods are proposed for CPSs modeled as finite-dimensional, discrete-time jump linear systems. In [176], a mode-independent state feedback controller is designed for CPSs subject to Markovian data dropouts. In [209] and [137], the two-mode-dependent state- and output-feedback stabilization

DOI: 10.1201/9781003260882-5

problems are investigated to stabilize CPSs with sensor-to-controller and controller-to-actuator channels delays modeled as two Markov chains, respectively.

In this chapter, the design problem of static output feedback delay compensation controller for CPSs with random network delay is considered, a delay compensation control approach is proposed to actively compensate the delay under Markovian jump system framework. Different from the delay compensation control approach proposed in Chapter 4, an output feedback strategy is used to generate the CIPs. The necessary and sufficient condition is proposed to perform the stability analysis of networked closed-loop system. Furthermore, by applying the singular Markovian jump system theory, the controller design problem is solved. Finally, the effectiveness of the proposed method is illustrated by using a numerical example.

5.2 SYSTEM DESCRIPTION

In this chapter, we consider the problem of output feedback delay compensation control for CPS shown in Figure 5.1.

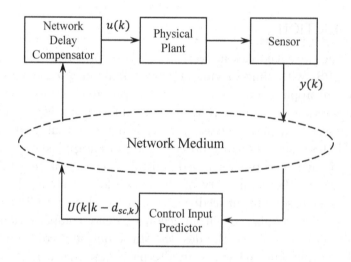

Figure 5.1 Block diagram of output-feedback networked delay compensation control

The dynamics of the remote-controlled plant is given by the following multi-input-multi-output (MIMO) discrete-time system:

$$x(k+1) = Ax(k) + Bu(k)$$
$$y(k) = Cx(k) \tag{5.1}$$

where $x(k) \in \mathbb{R}^n$ is the state vector; $u(k) \in \mathbb{R}^m$ is the control input; $y(k) \in \mathbb{R}^q$ is the output; and A, B, C are system matrices with appropriate dimensions.

The sensor is time-triggered and the system measurement is sampled periodically. Let $T_s > 0$ be the sampling period. At sampling instant k, the sampled measurement $y(k)$ and its time stamp i_k (that is, the time the measurement is sampled) are encapsulated into a packet and then sent to the remote controller via the network.

In practice, due to the limited bit rate of the communication channel and unavoidable errors or losses in the transmission, the data packets in CPSs usually suffer from network communication delays and data packet dropouts during network transmissions. Let d_k^{sc} and d_k^{ca} be the network communication delays in the sensor-to-controller and controller-to-actuator channels, respectively, and let $d_k = d_k^{sc} + d_k^{ca}$ be the round trip delay at time step k.

Assumption 5.1. *The random round trip delay d_k is assumed to be an integral multiple of the sampling period T_s, and at each time step k, d_k is assumed to be time-varying and bounded, that is, $0 < d_m \leq d_k \leq d_M$, where d_m and d_M are two positive integers.*

The objective of this chapter is to propose the static output feedback networked delay compensation control approach and controller design results for CPSs with random network communication delay.

5.3 STOCHASTIC DELAY COMPENSATION CONTROL SCHEME

The delay compensation controller proposed in this chapter is event-triggered. As long as there is a measurement packet reaching the controller across the network from the sensor, the controller calculates the control input signal immediately. When there is a random communication delay d_k^{sc} in the sensor-controller channel, the delay compensation controller is computed based on the received measurement $y(k - d_k^{sc})$, and the CIP is constructed as

$$U(k|k - d_k^{sc}) = \begin{bmatrix} u(k - d_k^{sc}|k - d_k^{sc}) \\ u(k - d_k^{sc} + 1|k - d_k^{sc}) \\ \vdots \\ u(k - d_k^{sc} + d_M|k - d_k^{sc}) \end{bmatrix}$$

where $u(k - d_k^{sc} + i|k - d_k^{sc}) = K_i y(k - d_k^{sc})$, $i = 0, 1, 2, \ldots, d_M$, with K_i, $i = 0, 1, 2, \ldots, d_M$, are the controller gains to be designed.

After finishing the calculations, the CIPs and the time stamp of the used output measurement are encapsulated into a packet and sent to the actuator via the network. The network delay compensator has the capability to compare the time stamps of the arrived control signal and the one stored in it, and will be updated only when the arrived control input is more recent than the existing one. Therefore, the network delay compensator is configured to accept the arrived packet only if its time stamp value is greater than that of the packet which the network delay compensator currently stores. When the data packet is received by the network delay compensator, we will select the control input $u(k|k - d_k)$ to actively compensate the random network communication delay, data packet dropout in CPSs.

With the above description in mind, the networked control system can be modeled as a linear discrete-time system with input delay d_k:

$$x(k + 1) = Ax(k) + Bu(k|k - d_k)$$
$$= Ax(k) + BK_{d_k}Cx(k - d_k) \qquad (5.2)$$

where d_k is the round trip delay at time step k. Here, we have omitted the sampling period T_s for brevity.

Remark 5.1. *It is noted that, in the proposed model (5.2), the random network delay is considered more precisely. More specifically, at the controller side, the sensor-to-controller channel delay d_k^{sc} is used to construct the control input packet $U(k|k - d_k^{sc})$, whereas at the actuator, the delay d_k^{ca} is utilized to select appropriate control input. Moreover, the controller can be seen as mode-dependent, since the gains K_i, $i = 0, 1, 2, \ldots, d_M$, are designed for different round trip delay d_k.*

In CPSs, it is reasonable to model the round trip delay $\{d_0, d_1, d_2, \ldots\}$ as a homogeneous ergodic Markov chain [209, 174], which takes values in $\mathcal{S} \triangleq \{d_m, d_m + 1, \ldots, d_M\}$, and $d_M = d_m + (M - m)$. Define an index set $\mathcal{I} \triangleq \{1, 2, \ldots, M - m + 1\}$ and an indexing function $\mathbb{I} : \mathcal{S} \to \mathcal{I}$ such that $\mathbb{I}(d_m + (i - 1)) = i$ which describes the ith mode of the delay. The transition probability matrix for the Markov chain is given by $\Lambda = [\lambda_{ij}]$, $i, j \in \mathcal{S}$. That is, from d_k transits to d_{k+1}, the delay jumps from mode i to j with probability λ_{ij}, that is, $\lambda_{ij} = \Pr(d_{k+1} = d_m + (j - 1)|d_k = d_m + (i - 1))$, where $\lambda_{ij} \geq 0$ and $\sum_{j \in \mathcal{S}} \lambda_{ij} = 1$ for all $i, j \in \mathcal{S}$.

Remark 5.2. *It is usually assumed that the network delay compensator always uses the most recent data, that is, the received packet with the largest time stamp value. If, at sampling time k, $U(k|k - d_k^{sc})$ is available, and its time stamp value is i_k, thus we have $i_k \leq k$ and time delay is $d_k = k - i_k$. Thus, at time $k + 1$, we have $i_k \leq i_{k+1} \leq k + 1$, and $d_k \geq k + 1 - i_{k+1} - 1 = d_{k+1} - 1$. In other words, we have $d_{k+1} \leq d_k + 1$, which indicates that $\Pr(d_{k+1} \geq d_k + 1) = 0$, $\lambda_{ij} = 0$ if $j \geq i + 1$ [209]. When $d_{k+1} = d_k + 1$ implies that data packet dropout occurs, in this case we can use $u(k + 1|k - d_k^{sc})$ in packet $U(k|k - d_k^{sc})$.*

5.4 STOCHASTIC STABILITY ANALYSIS

To model the CPS, we introduce the following augmented state vector:

$$\bar{x}(k) = \begin{bmatrix} x^T(k) & x^T(k - 1) & \cdots & x^T(k - d_M) \end{bmatrix}^T.$$

Thus, when there is a network delay d_k taking in $\mathcal{S}$ randomly, system (5.2) can be augmented as

$$\bar{x}(k + 1) = \bar{A}_{d_k} \bar{x}(k), \ d_k \in \mathcal{S} \tag{5.3}$$

where

$$\bar{A}_{d_k} = \left[\begin{array}{cccc} A & 0_{n \times (d_k - 1)n} & BK_{d_k}C & 0_{n \times (d_M - d_k)n} \\ \hline I_{d_M n \times d_M n} & & & 0_{d_M n \times n} \end{array} \right].$$

It can be seen that the closed-loop system in (5.3) is a delay-free Markovian jump linear system.

Before proceeding, we recall the following lemmas which will be needed in the deduction of the main results in this chapter.

Lemma 5.1. [37, 82] *Given a symmetric matrix Ξ and two matrices Γ and Π. Then, there exists a matrix Θ satisfying $\Xi + \Gamma\Theta\Pi + (\Gamma\Theta\Pi)^T < 0$ if and only if $\Gamma^{\perp}\Xi\Gamma^{\perp T} < 0$, $\Pi^{T\perp}\Xi\Pi^{T\perp T} < 0$.*

Lemma 5.2. [84] *System (5.3) is stochastically stable if and only if there exist matrices $Q_i > 0$, $i \in \mathcal{S}$, such that*

$$\bar{A}_i^T \bar{Q}_i \bar{A}_i - Q_i < 0, \ i \in \mathcal{S}, \tag{5.4}$$

where $\bar{Q}_i = \sum_{j \in \mathcal{S}} \lambda_{ij} Q_j$.

Now, we will give a necessary and sufficient condition on the stochastic stability of system (5.3), which will play an instrumental role in the controller design.

Theorem 5.1. *System (5.3) is stochastically stable if and only if there exist matrices $Q_i > 0$, $i \in \mathcal{S}$, matrices G_i and F_i satisfying the following coupled LMIs:*

$$\Upsilon_i \triangleq \begin{bmatrix} \Omega_i & \bar{A}_i^T F_i - G_i^T \\ * & \bar{Q}_i - F_i - F_i^T \end{bmatrix} < 0, \ i \in \mathcal{S}, \tag{5.5}$$

where $\bar{Q}_i = \sum_{j \in \mathcal{S}} \lambda_{ij} Q_j$ and $\Omega_i = \bar{A}_i^T G_i + G_i^T \bar{A}_i - Q_i$.

Proof. In order to prove the result, we only need to show the equivalence between (5.4) and (5.5).

"(5.4) $\Rightarrow$ (5.5)": To this end, we choose the following matrix variables in Lemma 5.1:

$$\Xi_i = \begin{bmatrix} -\bar{Q}_i & 0 & \bar{Q}_i \\ 0 & -Q_i & 0 \\ \bar{Q}_i & 0 & 0 \end{bmatrix}, \Gamma_i = \begin{bmatrix} 0 \\ \bar{A}_i^T \\ -I \end{bmatrix}, \Pi = \begin{bmatrix} 0 & 0 & I \\ 0 & I & 0 \end{bmatrix},$$

$$\Gamma_i^{\perp} = \begin{bmatrix} 0 & I & \bar{A}_i^T \\ I & 0 & 0 \end{bmatrix}, \Pi^{T\perp} = \begin{bmatrix} I & 0 & 0 \end{bmatrix}, \Theta_i = \begin{bmatrix} F_i & G_i \end{bmatrix}.$$

On the one hand, it can be verified that

$$\Gamma_i^{\perp}\Xi_i\Gamma_i^{\perp T} = \begin{bmatrix} 0 & I & \bar{A}_i^T \\ I & 0 & 0 \end{bmatrix} \begin{bmatrix} -\bar{Q}_i & 0 & \bar{Q}_i \\ 0 & -Q_i & 0 \\ \bar{Q}_i & 0 & 0 \end{bmatrix} \begin{bmatrix} 0 & I \\ I & 0 \\ \bar{A}_i & 0 \end{bmatrix}$$

$$= \begin{bmatrix} -Q_i & \bar{A}_i^T \bar{Q}_i \\ \bar{Q}_i \bar{A}_i & -\bar{Q}_i \end{bmatrix}. \tag{5.6}$$

It follows from Schur complement lemma that $\bar{A}_i^T \bar{Q}_i \bar{A}_i - Q_i < 0$ implies $\Gamma_i^{\perp}\Xi_i\Gamma_i^{\perp T} < 0$.

On the other hand, we can get

$$\Pi^{T\perp}\Xi_i\Pi^{T\perp T} = \begin{bmatrix} I & 0 & 0 \end{bmatrix} \begin{bmatrix} -\bar{Q}_i & 0 & \bar{Q}_i \\ 0 & -Q_i & 0 \\ \bar{Q}_i & 0 & 0 \end{bmatrix} \begin{bmatrix} I \\ 0 \\ 0 \end{bmatrix}$$

$$= -\bar{Q}_i < 0. \tag{5.7}$$

Now, it follows from Lemma 5.1 that (5.6) and (5.7) are equivalent to

$$\Xi_i + \Gamma_i\Theta_i\Pi + (\Gamma_i\Theta_i\Pi)^T = \begin{bmatrix} -\bar{Q}_i & 0 & \bar{Q}_i \\ * & \Omega_i & \bar{A}_i^T F_i - G_i^T \\ * & * & -F_i - F_i^T \end{bmatrix} < 0,$$

which is equivalent to (5.5) according to Schur complement lemma.

"(5.5) $\Rightarrow$ (5.4)": According to Lemma 5.1, we can conclude that $\Xi_i + \Gamma_i\Theta_i\Pi + (\Gamma_i\Theta_i\Pi)^T < 0$ implies that $\Gamma_i^\perp \Xi_i \Gamma_i^{\perp T} < 0$, which means (5.4) holds. This completes the proof. $\qquad\square$

Remark 5.3. *Compared with Lemma 5.2, the advantages associated with Theorem 5.1 are twofold. On the one hand, with the introduction of new additional matrices G_i and F_i, we obtain coupled LMIs in which the Lyapunov matrix Q_i is not involved in any product with the dynamic matrix $\bar{A}_i$. On the other hand, the conditions in Theorem 5.1 are with reduced conservatism due to the presence of the extra degrees of freedom provided by the introduction of matrices G_i and F_i. However, although the conservatism in stability results will be reduced by introducing slack variables, compared with Lemma 5.2, the computational complexity of Theorem 5.1 will be increased. For each LMI in (5.5), the number of the variables to be determined is $\frac{5(d_M+1)^2 n^2 + (d_M+1)n}{2}$. Nevertheless, when performing timing numerical experiments, the reduction of the conservatism far outweighs the increased complexity.*

It should be pointed out that in Theorem 5.1, we assume the transition probabilities λ_{ij}, $i, j \in \mathcal{S}$, are completely known, however, it may be difficult to determine precisely them in practice. Therefore, it is reasonable to assume that λ_{ij} belongs to a interval, say $\lambda_{ij} \in [\underline{\lambda}_{ij}, \overline{\lambda}_{ij}]$, where $0 \leq \underline{\lambda}_{ij}$ and $\overline{\lambda}_{ij} \leq 1$ are known parameters. With this observation, we have the following result.

Corollary 5.1. *System (5.3) with partially known transition probabilities $\lambda_{ij} \in [\underline{\lambda}_{ij}, \overline{\lambda}_{ij}]$, $i, j \in \mathcal{S}$, is stochastically stable if there exist matrices $Q_i > 0$, $i \in \mathcal{S}$, matrices G_i and F_i satisfying the following coupled LMIs:*

$$\begin{bmatrix} \overline{\Omega}_i & \bar{A}_i^T F_i - G_i^T \\ * & \overline{\mathcal{Q}}_i - F_i - F_i^T \end{bmatrix} < 0, \ i \in \mathcal{S}, \tag{5.8}$$

where $\overline{\mathcal{Q}}_i = \sum_{j\in\mathcal{S}} \overline{\lambda}_{ij} Q_j$, $\underline{\mathcal{Q}}_i = \sum_{j\in\mathcal{S}} \underline{\lambda}_{ij} Q_i$, $\overline{\Omega}_i = \bar{A}_i^T G_i + G_i^T \bar{A}_i - \underline{\mathcal{Q}}_i$.

Proof. Consider $\sum_{j\in\mathcal{S}} \lambda_{ij} = 1$ and $0 \leq \underline{\lambda}_{ij} \leq \lambda_{ij} \leq \overline{\lambda}_{ij} \leq 1$ for any $Q_j > 0$, the following two facts are true,

$$\sum_{j\in\mathcal{S}} \lambda_{ij} Q_j \leq \sum_{j\in\mathcal{S}} \overline{\lambda}_{ij} Q_j - \sum_{j\in\mathcal{S}} \lambda_{ij} Q_i \leq -\sum_{j\in\mathcal{S}} \underline{\lambda}_{ij} Q_i$$

then, it follows (5.4) that system (5.3) is stochastically stable if there exist matrices $Q_i > 0$, $i \in \mathcal{S}$, such that

$$\sum_{j\in\mathcal{S}} \overline{\lambda}_{ij} \bar{A}_i^T Q_j \bar{A}_i - \sum_{j\in\mathcal{S}} \underline{\lambda}_{ij} Q_i < 0. \tag{5.9}$$

By choosing the following matrix variables Ξ_i in Lemma 5.1:

$$\Xi_i = \begin{bmatrix} -\overline{\mathcal{Q}}_i & 0 & \overline{\mathcal{Q}}_i \\ 0 & -\underline{\mathcal{Q}}_i & 0 \\ \overline{\mathcal{Q}}_i & 0 & 0 \end{bmatrix},$$

and Γ_i, Π, $\Gamma_i^\perp$, $\Pi^{T\perp}$, and Θ_i are taken as same as the ones in the proof of Theorem 5.1, and following a similar line, we can show that (5.9) is equivalent to (5.8), and this completes the proof. □

Furthermore, it is worth mentioning that when the transition probabilities are completely unknown, system (5.3) becomes a switched linear system under arbitrary switching [172, 210], thus we have the following result.

Corollary 5.2. *System (5.3) with unknown transition probabilities λ_{ij}, $i, j \in \mathcal{S}$, is asymptotically stable if there exist matrices $Q_i > 0$, $i \in \mathcal{S}$, G_i and F_i satisfying the following LMIs:*

$$\begin{bmatrix} \Omega_i & \bar{A}_i^T F_i - G_i^T \\ * & Q_j - F_i - F_i^T \end{bmatrix} < 0, \ i, j \in \mathcal{S}, \tag{5.10}$$

where Ω_i is given in Theorem 5.1.

Proof. In Theorem 5.1, it has been shown that system (5.3) is stochastically stable if and only if (5.5) holds. Thus, owing to $\sum_{j \in \mathcal{S}} \lambda_{ij} = 1$, we can rewrite the left-hand side of (5.5) as

$$\Upsilon_i = \sum_{j \in \mathcal{S}} \lambda_{ij} \begin{bmatrix} \Omega_i & \bar{A}_i^T F_i - G_i^T \\ F_i^T \bar{A}_i - G_i & Q_j - F_i - F_i^T \end{bmatrix}.$$

Then, since $\lambda_{ij} \geq 0$, $i, j \in \mathcal{S}$, it is straightforward that (5.5) holds if (5.10) holds, which completes the proof. □

Remark 5.4. *Compared with the stability results for switched linear systems, such as [172], the conditions in Corollary 5.2 are also a new sufficient conditions to guarantee the switched linear system is globally uniformly asymptotically stable in the discrete-time framework.*

5.5 CONTROLLER SYNTHESIS

It can be seen that K_{d_k} in (5.3) is embedded in the middle of matrix B and C, thus it is difficult to give a parametrized solution. The fundamental idea is to extract K_{d_k} from the middle of the two matrices. Similar to [138], we choose the the following augmented state variable:

$$\tilde{x}(k) \triangleq \begin{bmatrix} \bar{x}(k) \\ u(k|k - d_k) \end{bmatrix}$$

as a new state variable. Then the closed-loop system (5.3) can be rewritten as the following augmented system:

$$E\tilde{x}(k+1) = \tilde{A}_{d_k}\tilde{x}(k), \tag{5.11}$$

where

$$E = \left[\begin{array}{c|c} I_{(d_M+1)n\times(d_M+1)n} & 0 \\ \hline 0 & 0 \end{array}\right], \tilde{A}_{d_k} = \left[\begin{array}{c|c} \bar{A} & \bar{B} \\ \hline K_{d_k}\bar{C}_{d_k} & -I_{m\times m} \end{array}\right]$$

$$\bar{A} = \left[\begin{array}{cc} A & 0_{n\times d_M n} \\ \hline I_{d_M n\times d_M n} & 0_{d_M n\times n} \end{array}\right], \bar{B} = \left[\begin{array}{c} B \\ 0_{d_M n\times m} \end{array}\right],$$

$$\bar{C}_{d_k} = \left[\begin{array}{ccc} 0_{q\times d_k n} & C & 0_{q\times(d_M-d_k)n} \end{array}\right].$$

Thus, system (5.11) is a discrete-time singular Markovian jump system.

The following definition and lemma will be used in the derivation of the main results.

Definition 5.1. *[178, 173]*

1) *The discrete-time singular Markovian jump system in (5.11) is said to be regular if, for each $i \in \mathcal{S}$, $\det(zE - \tilde{A}_i)$ is not identically zero.*

2) *The discrete-time singular Markovian jump system in (5.11) is said to be causal if, for each $i \in \mathcal{S}$, $\deg(\det(zE - \tilde{A}_i)) = \text{rank}(E)$.*

3) *The discrete-time singular Markovian jump system in (5.11) is said to be stochastically stable if for any $x_0 \in \mathbb{R}^n$ and $r_0 \in \mathcal{S}$, there exists a scalar $M(x_0, r_0) > 0$ such that*

$$\lim_{N\to\infty} \mathbb{E}\left\{\sum_{k=0}^{N} \|\tilde{x}(k,x_0,r_0)\|^2 | x_0, r_0\right\} \leq M(x_0, r_0),$$

where $\tilde{x}(k,x_0,r_0)$ denote the solution to system (5.11) at time k under the initial conditions x_0 and r_0.

4) *The discrete-time singular Markovian jump system in (5.11) is said to be stochastically admissible if it is regular, causal, and stochastically stable.*

Lemma 5.3. *[178] The discrete-time singular Markovian jump system in (5.11) is stochastically admissible if and only if there exist symmetric matrices P_i, $i \in \mathcal{S}$, such that the following coupled LMIs hold for each $i \in \mathcal{S}$:*

$$E^T P_i E \geq 0, \tag{5.12}$$
$$\tilde{A}_i^T \bar{P}_i \tilde{A}_i - E^T P_i E < 0, \tag{5.13}$$

where $\bar{P}_i = \sum_{j\in\mathcal{S}} \lambda_{ij} P_j$.

Now, we have the following lemma, which plays an important role in deriving the output controller.

Lemma 5.4. *The discrete-time singular Markovian jump system (5.11) is stochastically admissible if and only if there exist matrices $P_i > 0$, $i \in \mathcal{S}$ and a symmetric and nonsingular matrix Φ, such that the following coupled LMIs hold for each $i \in \mathcal{S}$*

$$\tilde{A}_i^T (\bar{P}_i - R^T \Phi R) \tilde{A}_i - E^T P_i E < 0, \tag{5.14}$$

where $R \in \mathbb{R}^{m \times ((d_M + 1)n + m)}$ is a matrix with the properties that $RE = 0$ with $rank(R) = m$, and $\bar{P}_i = \sum_{j \in \mathcal{S}} \lambda_{ij} P_j$.

Proof. Sufficiency. Let $Y_i = P_i - R^T \Phi R$ in (5.12), we can get

$$E^T Y_i E = E^T (P_i - R^T \Phi R) E = E^T P_i E \geq 0, \tag{5.15}$$

$$\tilde{A}_i^T \bar{Y}_i \tilde{A}_i - E^T Y_i E < 0, \tag{5.16}$$

where $\bar{Y}_i = \sum_{j \in \mathcal{S}} \lambda_{ij} Y_j$.

Necessity. Suppose that the discrete-time singular Markovian jump system (5.11) is stochastically admissible. Now, noticing the partition of $\tilde{A}_i$ in (5.11), and define that

$$\tilde{A}_1 = \bar{A}, \ \tilde{A}_2 = \bar{B}, \ \tilde{A}_{3i} = K_i \bar{C}_i, \ \tilde{A}_4 = -I_{m \times m}. \tag{5.17}$$

Then, select a nonsingular matrix as $\mathcal{L}_i = \begin{bmatrix} I & \tilde{A}_{3i}^T \\ 0 & I \end{bmatrix}$, and it can be verified that

$$\check{A}_i = \tilde{A}_i \mathcal{L}_i^T = \begin{bmatrix} \check{A}_{1i} & \tilde{A}_2 \\ 0 & \tilde{A}_4 \end{bmatrix}, \tag{5.18}$$

where $\check{A}_{1i} = \tilde{A}_1 + \tilde{A}_2 \tilde{A}_{3i} = \bar{A} + \bar{B} K_i \bar{C}_i$.

Therefore, the stochastic stability of system (5.11) implies that the stochastic stability of the following discrete Markovian jump system $\xi(k+1) = \check{A}_{1d_k} \xi(k)$, where $\xi(k)$ is an auxiliary variable with appropriate dimension, $\check{A}_{1d_k} = \bar{A} + \bar{B} K_{d_k} \bar{C}_{d_k}$. It follows that there exist matrices $\check{P}(i) > 0$, $i \in \mathcal{S}$, such that $\check{A}_{1i}^T \bar{\check{P}}_i \check{A}_{1i} - \check{P}_i < 0$, where $\bar{\check{P}}_i = \sum_{j \in \mathcal{S}} \lambda_{ij} \check{P}_j$.

Therefore, we can always find a sufficiently large scalars $\kappa > 0$ such that, for $i \in \mathcal{S}$,

$$\check{A}_i^T \begin{bmatrix} \bar{\check{P}}_i & 0 \\ 0 & \kappa I \end{bmatrix} \check{A}_i - E^T \begin{bmatrix} \check{P}_i & 0 \\ 0 & \kappa I \end{bmatrix} E - \begin{bmatrix} 0 \\ \tilde{A}_4^T \end{bmatrix} 2\kappa I \begin{bmatrix} 0 & \tilde{A}_4 \end{bmatrix}$$

$$= \begin{bmatrix} \Sigma_{(1,1)} & \Sigma_{(1,2)} \\ \Sigma_{(1,2)}^T & \Sigma_{(2,2)} \end{bmatrix} < 0,$$

where $\Sigma_{(1,1)} = \check{A}_{1i}^T \bar{\check{P}}_i \check{A}_{1i} - \check{P}_i$, $\Sigma_{(1,2)} = \check{A}_{1i}^T \bar{\check{P}}_i \tilde{A}_2$, $\Sigma_{(2,2)} = \tilde{A}_2^T \bar{\check{P}}_i \tilde{A}_2 - \kappa I$.

Let $\mathcal{I} = [\ 0\ \ I\ ]$, we obtain $E\mathcal{L}_i^T = E$, $\mathcal{I}\tilde{A}_i\mathcal{L}_i^T = \begin{bmatrix} 0 & \tilde{A}_4 \end{bmatrix}$.

Now, define $P_i = \begin{bmatrix} \check{P}_i & 0 \\ 0 & \kappa I \end{bmatrix}$, $R = W\mathcal{I}$, $\Phi = 2\kappa W^{-T}W^{-1}$, where W is an nonsingular matrix with appropriate dimension. Then, we can get $\mathcal{L}_i \left(\tilde{A}_i^T \bar{P}_i \tilde{A}_i - E^T P_i E - \tilde{A}_i^T R^T \Phi R \tilde{A}_i \right) \mathcal{L}_i^T < 0$, which is equivalent to (5.14). This completes the proof. $\qquad\square$

Based on Theorem 5.1 and Lemma 5.4, we can obtain the following corollary immediately.

Corollary 5.3. *Singular Markovian jump system (5.3) is stochastically admissible if and only if there exist matrices $Q_i > 0$, $i \in \mathcal{S}$, G_i, F_i, and Φ satisfying the following LMIs:*

$$\begin{bmatrix} \Theta_i & \tilde{A}_i^T F_i - G_i^T \\ * & \bar{Q}_i - F_i - F_i^T - R^T \Phi R \end{bmatrix} < 0, \ i \in \mathcal{S},$$

where $\bar{Q}_i = \sum_{j \in \mathcal{S}} \lambda_{ij} Q_j$ and $\Theta_i = \tilde{A}_i^T G_i + G_i^T \tilde{A}_i - E^T Q_i E$.

Proof. The proof is similar to Theorem 5.1 with Ξ_i re-defined as

$$\Xi_i = \begin{bmatrix} -\bar{Q}_i & 0 & \bar{Q}_i \\ 0 & -E^T Q_i E & 0 \\ \bar{Q}_i & 0 & -R^T \Phi R \end{bmatrix},$$

and the details are omitted here. $\qquad\square$

Remark 5.5. *In view of Remark 5.3, Corollary 5.3 is also an alternative necessary and sufficient condition to guarantee that the singular Markovian jump linear system is stochastically admissible.*

With Corollary 5.1 and Lemma 5.4, we can obtain the following result for the partially known transition probabilities case.

Corollary 5.4. *Singular Markovian jump system (5.3) with partially known transition probabilities $\lambda_{ij} \in [\underline{\lambda}_{ij}, \overline{\lambda}_{ij}]$, $i, j \in \mathcal{S}$, is stochastically admissible if there exist matrices $Q_i > 0$, $i \in \mathcal{S}$, G_i, F_i, and Φ satisfying the following LMIs:*

$$\begin{bmatrix} \Theta_i & \tilde{A}_i^T F_i - G_i^T \\ * & \overline{\mathcal{Q}}_i - F_i - F_i^T - R^T \Phi R \end{bmatrix} < 0, \ i \in \mathcal{S},$$

where $\overline{\mathcal{Q}}_i = \sum_{j \in \mathcal{S}} \overline{\lambda}_{ij} Q_j$, $\underline{\mathcal{Q}}_i = \sum_{j \in \mathcal{S}} \underline{\lambda}_{ij} Q_i$ and $\Theta_i = \tilde{A}_i^T G_i + G_i^T \tilde{A}_i - E^T \underline{\mathcal{Q}}_i E$.

Proof. The proof is similar to Corollary 5.1 with Ξ_i re-defined as

$$\Xi_i = \begin{bmatrix} -\overline{\mathcal{Q}}_i & 0 & \overline{\mathcal{Q}}_i \\ 0 & -E^T \underline{\mathcal{Q}}_i E & 0 \\ \overline{\mathcal{Q}}_i & 0 & -R^T \Phi R \end{bmatrix},$$

and the details are omitted here. $\qquad\square$

Now, we are ready to give our main result on the solution of the controller design problem.

Theorem 5.2. *For given tuning scalars α_i, and β_i, system (5.3) is stochastically admissible if there exist matrices $Q_i > 0$, $i \in \mathcal{S}$, G_i, F_i, L_i, and Φ satisfying the following LMIs:*

$$\begin{bmatrix} \Xi_i & \Phi_i^T - G_i^T \\ * & \bar{Q}_i - F_i - F_i^T - R^T \Phi R \end{bmatrix} < 0, \ i \in \mathcal{S}, \tag{5.19}$$

where $\bar{Q}_i = \sum_{j \in \mathcal{S}} \lambda_{ij} Q_j$, $\Xi_i = \Psi_i + \Psi_i^T - E^T Q_i E$ and

$$\Phi_i = \begin{bmatrix} F_{1i}^T \bar{A} + \alpha_i \mathcal{I}^T L_i \bar{C}_i & F_{1i}^T \bar{B} - \alpha_i \mathcal{I}^T J_i^T \\ F_{2i}^T \bar{A} + L_i \bar{C}_i & F_{2i}^T \bar{B} - J_i^T \end{bmatrix},$$

$$\Psi_i = \begin{bmatrix} G_{1i}^T \bar{A} + \beta_i \mathcal{I}^T L_i \bar{C}_i & G_{1i}^T \bar{B} - \beta_i \mathcal{I}^T J_i^T \\ G_{2i}^T \bar{A} + L_i \bar{C}_i & G_{2i}^T \bar{B} - J_i^T \end{bmatrix},$$

with $\mathcal{I} = [I_m, \ 0]$ and $F_i = \begin{bmatrix} F_{1i} & F_{2i} \\ \alpha_i J_i \mathcal{I} & J_i \end{bmatrix}$, $G_i = \begin{bmatrix} G_{1i} & G_{2i} \\ \beta_i J_i \mathcal{I} & J_i \end{bmatrix}$. Furthermore, if (5.19) is feasible, an output feedback controller gain K_i is given by

$$K_i = J_i^{-T} L_i. \tag{5.20}$$

Proof. The desired results can be obtained by using Corollary 5.3 and noticing the structures of $\mathcal{I}$, F_i, and G_i immediately, thus the details are omitted. $\square$

Remark 5.6. *In Theorem 5.2, the tuning scalars α_i, and β_i are introduced to reduce conservatism. To find the tuning scalars, we first select a set of initial tuning parameters and such that the LMIs are feasible, then, apply a numerical optimization algorithm, such as the program* fminsearch *in the Optimization Toolbox of Matlab [34, 221].*

Remark 5.7. *In [173], the controller design is carried out under the assumption that the output matrix C is of full row rank, which will limit the applicability of the results. However, in this chapter, this assumption is removed.*

For the partially known transition probabilities case, we have the following result.

Corollary 5.5. *For given tuning scalars α_i and β_i, system (5.3) with partially known transition probabilities $\lambda_{ij} \in [\underline{\lambda}_{ij}, \overline{\lambda}_{ij}]$, $i, j \in \mathcal{S}$, is stochastically admissible if there exist matrices $Q_i > 0$, $i \in \mathcal{S}$, G_i, F_i, L_i and Φ satisfying the following LMIs:*

$$\begin{bmatrix} \overline{\Xi}_i & \Phi_i^T - G_i^T \\ * & \overline{\mathcal{Q}}_i - F_i - F_i^T - R^T \Phi R \end{bmatrix} < 0, \ i, j \in \mathcal{S}, \tag{5.21}$$

where $\overline{\mathcal{Q}}_i = \sum_{j \in \mathcal{S}} \overline{\lambda}_{ij} Q_j$, $\underline{\mathcal{Q}}_i = \sum_{j \in \mathcal{S}} \underline{\lambda}_{ij} Q_i$, $\overline{\Xi}_i = \Psi_i + \Psi_i^T - E^T \underline{\mathcal{Q}}_i E$, Φ_i, and Ψ_i are given in Theorem 5.2. Furthermore, if (5.21) is feasible, an output feedback controller gain K_i is given by (5.20).

For the completely unknown transition probabilities case, we have the following result.

Corollary 5.6. *For given tuning scalars α_i, and β_i, system (5.3) with unknown transition probabilities λ_{ij}, $i,j \in \mathcal{S}$, is asymptotically stable if there exist matrices $Q_i > 0$, $i \in \mathcal{S}$, G_i, F_i, L_i and Φ satisfying the following LMIs:*

$$\begin{bmatrix} \Xi_i & \Phi_i^T - G_i^T \\ * & Q_j - F_i - F_i^T - R^T\Phi R \end{bmatrix} < 0, \ i,j \in \mathcal{S}, \tag{5.22}$$

where Ξ_i, Φ_i and Ψ_i are given in Theorem 5.2. Furthermore, if (5.22) is feasible, an output feedback controller gain K_i is given by (5.20).

5.6 NUMERICAL SIMULATION

In this section, we will illustrate the developed theory via a numerical example and simulation. Let the controlled plant be a servo motor control system which consists of a DC motor, load plate, speed, and angle sensors. The model of the motor control plant at sampling period 0.04s was identified to be [228]

$$G(z^{-1}) = \frac{A(z^{-1})}{B(z^{-1})} = \frac{0.05409z^{-2} + 0.115z^{-3} + 0.0001z^{-4}}{1 - 1.12z^{-1} - 0.213z^{-2} + 0.335z^{-3}}.$$

This system can also be written in state-space form (5.1) with the following system matrices:

$$A = \begin{bmatrix} 1.12 & 0.213 & -0.335 \\ 1 & 0 & 0 \\ 0 & 1 & 0 \end{bmatrix}, B = \begin{bmatrix} 1 \\ 0 \\ 0 \end{bmatrix}, C = \begin{bmatrix} 0.0541 & 0.1150 & 0.0001 \end{bmatrix}.$$

Now, we suppose that the plant is controlled over network, and our purpose is to design output feedback delay compensation controllers such that the closed-loop system is stochastically stable for all admissible random network communication delays.

We first assume the transition probabilities λ_{ij}, $i,j \in \mathcal{S}$, are completely known, and the transition probability matrix is

$$\Lambda = \begin{bmatrix} 0.2 & 0.8 & 0.0 \\ 0.2 & 0.6 & 0.2 \\ 0.4 & 0.5 & 0.1 \end{bmatrix}.$$

Taking $\alpha_1 = -0.9$, $\alpha_2 = -0.5$, $\alpha_3 = 0.2$, $\beta_i = 1$, $i = 1, 2, 3$, by solving the feasibility problem in Theorem 5.2, we can find a set of controller gains as $K_1 = -1.5980$, $K_2 = -1.2802$, $K_3 = 0.0032$. When there is no delay compensation, we can obtain the controller gain as $K = -1.1620$.

Simulations are carried out by applying the above obtained output-feedback delay compensation controllers to the physical plant. The initial condition is assumed to be $[0, 1, 0]^T$. In this simulation, the network communication delay is generated randomly according to the transition probability matrix Λ, and the lower and upper bounds of network communication delay are assumed as $d_m = 1$ and $d_M = 3$. The state responses and control input are depicted in Figure 5.2, from which we can see that all three state components are stochastically stable with or without delay compensation, while it is easy to see that the proposed delay compensation control approach has better control performance, which illustrate the advantage of delay compensation control scheme proposed in this chapter.

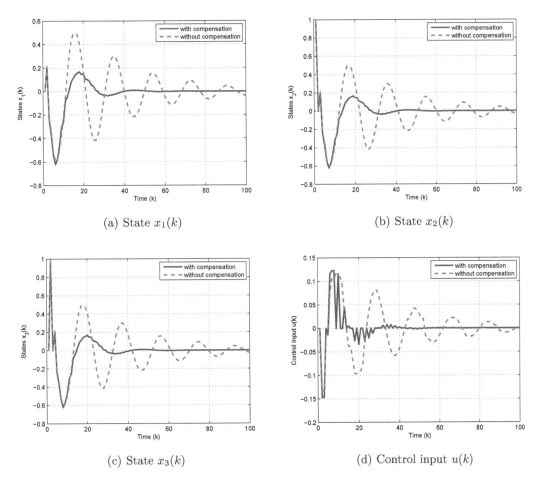

(a) State $x_1(k)$

(b) State $x_2(k)$

(c) State $x_3(k)$

(d) Control input $u(k)$

Figure 5.2 States and control input of motor control system (known transition probabilities case)

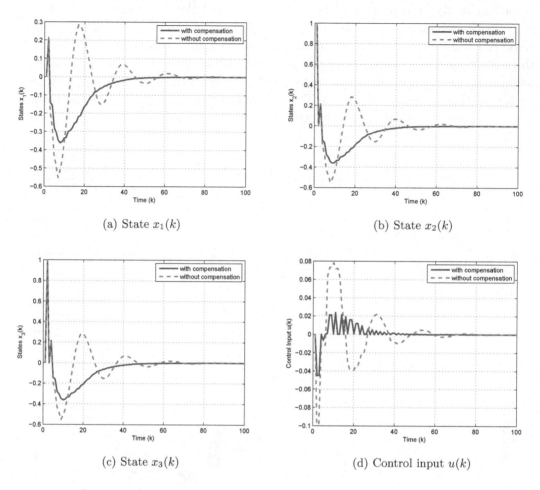

(a) State $x_1(k)$

(b) State $x_2(k)$

(c) State $x_3(k)$

(d) Control input $u(k)$

Figure 5.3 States and control input of motor control system (partially known transition probabilities case)

Next, we assume partially known transition probabilities $\lambda_{ij} \in [\underline{\lambda}_{ij}, \overline{\lambda}_{ij}]$, $i, j \in \mathcal{S}$, where $\underline{\lambda}_{ij}$ and $\overline{\lambda}_{ij}$ are taken as

$$\underline{\Lambda} = [\underline{\lambda}_{ij}] = \begin{bmatrix} 0.19 & 0.79 & 0 \\ 0.17 & 0.58 & 0.17 \\ 0.35 & 0.45 & 0.05 \end{bmatrix}, \overline{\Lambda} = [\overline{\lambda}_{ij}] = \begin{bmatrix} 0.21 & 0.81 & 0 \\ 0.21 & 0.61 & 0.21 \\ 0.45 & 0.55 & 0.15 \end{bmatrix},$$

then by selecting $\alpha_1 = -0.5$, $\alpha_2 = -1.2$, $\alpha_3 = -0.5$, $\beta_i = 1$, $i = 1, 2, 3$, we can obtain the controller gains as $K_1 = -0.0166$, $K_2 = -0.3925$, $K_3 = -0.0026$, and if there is no delay compensation, we can obtain the controller gain as $K = -0.8563$. The simulation results are shown in Figure 5.3, it can be seen that closed-loop system is stochastically stable, the proposed delay compensation control approach is also effective for the partially known transition probabilities case.

Finally, we assume the transition probabilities λ_{ij}, $i, j \in \mathcal{S}$, are completely unknown. By choosing $\alpha_1 = -1.0$, $\alpha_2 = 1.2$, $\alpha_3 = 1.5$, $\beta_1 = -1.0$, $\beta_2 = 0.2$, and $\beta_3 = -0.5$, according to Corollary 5.6, we can find delay compensation controller gains as $K_1 = -0.6847$, $K_2 = -0.3254$, $K_3 = -0.0687$. Furthermore, we obtain the controller gain as $K = -0.5851$ if no delay compensation is considered. The simulation is shown in Figure 5.4. It is observed that the delay compensation controller also works well when the transition probabilities are completely unknown, and outperforms the controller without compensation.

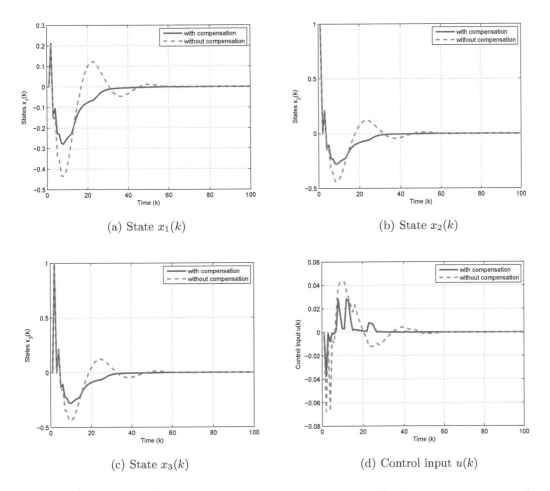

(a) State $x_1(k)$

(b) State $x_2(k)$

(c) State $x_3(k)$

(d) Control input $u(k)$

Figure 5.4 States and control input of motor control system (unknown transition probabilities case)

5.7 SUMMARY

In this chapter, the static output feedback networked delay compensation control approach has been proposed for CPSs with random network communication delays. Under Markovian jump linear system framework, the necessary and sufficient condition has been proposed to ensure the stochastic stability of networked closed-loop system. Then, the controller design problem has been solved by using the singular Markovian jump system theory. Finally, a numerical example has been used to show the effectiveness of the proposed method.

SUMMARY

CHAPTER 6

Fuzzy Networked Delay Compensation Control for Nonlinear CPSs

6.1 INTRODUCTION

In the previous chapters, the networked delay compensation control problems are proposed for linear physical plant in CPSs. In this chapter, our main objective is to develop the network delay compensation controllers for a class of nonlinear plant in CPSs with both sensor-to-controller and controller-to-actuator channel communication delays. With the fact that, the well-known Takagi-Sugeno (T-S) fuzzy model is a powerful tool in approximating complex nonlinear systems [30, 151, 203, 11], and then, fuzzy controllers are designed based on the parallel distributed compensation (PDC) technique. Using the T-S fuzzy model, some results on networked controller designs have recently been published [191, 218, 197, 128]. In [191], the problem of stability analysis and stabilization control design for T-S fuzzy systems with probabilistic interval delay is considered. The stabilization problem for networked stochastic systems with transmitted data dropout is investigated in [218], where the plant in the CPSs is a discrete stochastic time-delay nonlinear system represented by a T-S fuzzy model. In [197], the robust control problem is investigated for a class of uncertain nonlinear systems represented by a T-S fuzzy model with uncertainties, and both network communication delay and data dropout are addressed. In [128], the robust output-feedback control problem is considered for networked nonlinear systems with multiple packet dropouts, the nonlinear plant is represented by T-S fuzzy affine dynamic models, and stochastic variables that satisfy the Bernoulli random binary distribution are adopted to characterize the data dropout phenomenon. In [193], the event-triggered fuzzy H_∞ control and tracking control problems are studied for networked nonlinear systems by using the deviation bounds of asynchronous normalized member-ship functions. For the recent advances of fuzzy-model-based nonlinear systems, please refers to the recent survey paper [129].

This chapter is concerned with the network delay compensation control problem for nonlinear plant in CPSs, and the fuzzy network delay compensation control

DOI: 10.1201/9781003260882-6

approaches are proposed to actively compensate the network communication delay in the fuzzy control framework. The nonlinear plant is represented by a T-S fuzzy model, and the CIPs are constructed based on PDC technique. Both state and output feedback fuzzy delay compensation controllers are designed. Finally, two examples are provided to illustrate the effectiveness and applicability of the developed techniques.

6.2 SYSTEM DESCRIPTION

In this chapter, we consider the problem of fuzzy delay compensation control of the nonlinear plant in CPSs shown in Figure 6.1. It is well known that many nonlinear

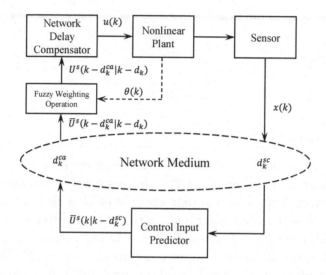

Figure 6.1 Block diagram of fuzzy networked delay compensation control

systems can be expressed as a set of linear systems in local operating regions. The nonlinear discrete-time plant can be represented by the T-S fuzzy model as follows:
Plant Rule i: IF $\theta_1(k)$ is M_{i1}, $\theta_2(k)$ is M_{i2} and $\cdots$ and $\theta_\omega(k)$ is $M_{i\omega}$ THEN

$$x(k+1) = A_i x(k) + B_i u(k),$$
$$y(k) = C_i x(k), \tag{6.1}$$

where $\theta_1(k)$, $\theta_2(k)$, $\cdots$ and $\theta_\omega(k)$ are the premise variables, and M_{ij}, $i = 1, 2, \cdots, r$, $j = 1, 2, \cdots, \omega$ are fuzzy sets, $x(k) \in \mathbb{R}^n$ is the state variable, $u(k) \in \mathbb{R}^m$ is the control input, $y(k) \in \mathbb{R}^\nu$ is the system output variable, and A_i, B_i, and C_i are known matrices. r is the number of IF-THEN rules. For the sake of notational convenience, we denote $\mathcal{N} \triangleq \{1, 2, \cdots, r\}$.

Given a pair $(x(k), u(k))$, the defuzzified output of the fuzzy system is inferred as follows:

$$x(k+1) = \sum_{i=1}^{r} h_i(\theta_k) \{A_i x(k) + B_i u(k)\},$$

$$y(k) = \sum_{i=1}^{r} h_i(\theta_k) C_i x(k), \qquad (6.2)$$

where $h_i(\theta(k)) = \frac{\omega_i(\theta(k))}{\sum_{i=1}^{r} \omega_i(\theta(k))}$, $\omega_i(\theta(k)) = \prod_{l=1}^{\omega} M_{il}(\theta_l(k))$, in which $M_{il}(\theta_l(k))$ is the membership degree of $\theta_l(k)$ in M_{il}, $i \in \mathcal{N}$. It is assumed that $\omega_i(\theta(k)) \geq 0, i \in \mathcal{N}$. Then, it can be seen that $\sum_{i=1}^{r} \omega_i(\theta(k)) > 0$, for all k. Therefore, for all k, $h_i(\theta(k)) \geq 0, i \in \mathcal{N}$, $\sum_{i=1}^{r} h_i(\theta_k) = 1$. In what follows, we will drop the argument of $h_i(\theta_k)$ for simplicity.

In Figure 6.1, the sensor is time-triggered and the system measurement is sampled periodically. At sampling instant k, the sampled measurement $x(k)$ (or $y(k)$) and its time stamp i_k (that is, the time the measurement is sampled) are encapsulated into a packet and then sent to the remote controller via the network.

Similar to the previous chapters, define d_k^{sc} and d_k^{ca} as the network communication delays in the sensor-to-controller and controller-to-actuator channel, respectively, and let $d_k = d_k^{sc} + d_k^{ca}$ be the round trip delay at time step k. As for round trip delay d_k, we make the following assumption.

Assumption 6.1. *At each time step k, the round trip delay d_k is time-varying and bounded, that is, $1 \leq d_m \leq d_k \leq d_M$, where d_m and d_M are two positive integers. For simplicity, we denote $\mathcal{S} \triangleq \{d_m, d_m + 1, \ldots, d_M\}$.*

The objective of this chapter is to propose the fuzzy delay compensation control schemes for the nonlinear plant in CPSs, and both the state and output feedback controllers are designed.

6.3 FUZZY DELAY COMPENSATION CONTROL SCHEME

In this section, we assume that the state is available for feedback control, and the sensor-to-controller channel network communication delay is d_k^{sc}, the CIP is constructed based on the received data $x(k - d_k^{sc})$ and the local state-based controller can be determined as

$$\bar{u}^s(k - d_k^{sc} + j | k - d_k^{sc}) = K_{ij} x(k - d_k^{sc}), i \in \mathcal{N}, j \in \mathcal{S}, \qquad (6.3)$$

where K_{ij}, $i \in \mathcal{N}$, $j \in \mathcal{S}$, are the controller gains to be designed, then the local state-based CIP can be constructed by

$$\overline{U}^s(k|k - d_k^{sc}) = \begin{bmatrix} \bar{u}^s(k - d_k^{sc}|k - d_k^{sc}) \\ \bar{u}^s(k - d_k^{sc} + 1|k - d_k^{sc}) \\ \vdots \\ \bar{u}^s(k - d_k^{sc} + d_M|k - d_k^{sc}) \end{bmatrix}. \qquad (6.4)$$

Otherwise, if only output variables are available for feedback control, the CIP is constructed based on the received data $y(k - d_k^{sc})$, where d_k^{sc} is the communication delay in the sensor-to-controller channel. The local output-based controller can be determined by

$$\bar{u}^o(k - d_k^{sc} + j|k - d_k^{sc}) = H_{ij} y(k - d_k^{sc}), i \in \mathcal{N}, \ j \in \mathcal{S}, \tag{6.5}$$

where $H_{ij}, i \in \mathcal{N}, j \in \mathcal{S}$, are the controller gains to be designed.

Then, the output-based CIP is constructed as

$$\overline{U}^o(k|k - d_k^{sc}) = \begin{bmatrix} \bar{u}^o(k - d_k^{sc}|k - d_k^{sc}) \\ \bar{u}^o(k - d_k^{sc} + 1|k - d_k^{sc}) \\ \vdots \\ \bar{u}^o(k - d_k^{sc} + d_M|k - d_k^{sc}) \end{bmatrix}. \tag{6.6}$$

After finishing the calculation, the state- or output-based CIP and the time stamp of the used plant output are encapsulated into a packet and sent to the fuzzy weighting operation side via the network, the defuzzified CIP $U^s(k - d_k^{ca}|k - d_k)$ or $U^o(k - d_k^{ca}|k - d_k)$ are computed by the parallel distributed compensation (PDC) technique [169, 168, 10, 113]. According to the network delay compensation strategy proposed in the previous chapters, if there is a new data packet stored in the network delay compensator after the comparison process, we will select the state feedback control input $u^s(k|k - d_k)$ in the data packet $U^s(k - d_k^{ca}|k - d_k)$ or the output feedback control input $u^o(k|k - d_k)$ in the data packet $U^o(k - d_k^{ca}|k - d_k)$ to actively compensate the network communication delay.

Remark 6.1. *Motivated by Huang and Nguang [78], the premise vector θ_k is supposed to be connected to the plant via point-to-point architecture, thus the plant and controller can share the same premise vector, and this case is also considered in existing results [78]. It is worth mentioning that, if the premise variables for the fuzzy delay compensator are difficult to determine, one can choose different premise variables between the plant and fuzzy delay compensator, that is the premise variables are asynchronous, in this case, T-S fuzzy controller based on PDC does not offer any advantages over a linear controller [112].*

With the state-based delay compensation controller $u^s(k|k - d_k)$, the closed-loop networked control system can be represented as:

$$x(k + 1) = \sum_{i=1}^{r} \sum_{j=1}^{r} h_i(\theta_k) h_j(\theta_k) \left\{ A_i x(k) + B_i K_{jd_k} x(k - d_k) \right\}, \tag{6.7}$$

where d_k is the round trip delay at time step k.

By using the output-based delay compensation controller $u^o(k|k - d_k)$, the closed-loop networked control system can be represented as:

$$x(k + 1) = \sum_{i=1}^{r} \sum_{j=1}^{r} h_i(\theta_k) h_j(\theta_k) \left\{ A_i x(k) + B_i H_{jd_k} y(k - d_k) \right\}. \tag{6.8}$$

6.4 STABILITY ANALYSIS AND CONTROLLER DESIGN

In this section, we will analyze the stability for closed-loop networked control system (6.7) and (6.8), respectively. We first examine the state feedback case. For simplicity, we rewrite system (6.7) into the following form,

$$x(k+1) = A_h x(k) + B_h K_{h,d_k} x(k - d_k), \quad (6.9)$$

where $A_h = \sum_{i=1}^{r} h_i(\theta_k) A_i$, $B_h = \sum_{i=1}^{r} h_i(\theta_k) B_i$, $K_{h,d_k} = \sum_{i=1}^{r} h_i(\theta_k) K_{id_k}$.

To analyze the stability of system (6.7), we introduce the following augmented state vector $\bar{x}(k) = \begin{bmatrix} x^T(k) & x^T(k-1) & \cdots & x^T(k-d_M) \end{bmatrix}^T$. Thus, when there is time-varying delay d_k taking in $\mathcal{S}$, system (6.9) can be augmented as

$$\bar{x}(k+1) = \bar{A}_{h,d_k} \bar{x}(k), \quad (6.10)$$

where

$$\bar{A}_{h,d_k} = \left[\begin{array}{cccc} A_h & 0_{n \times (d_k-1)n} & B_h K_{h,d_k} & 0_{n \times (d_M-d_k)n} \\ \hline I_{d_M n \times d_M n} & & 0_{d_M n \times n} \end{array} \right], d_k \in \mathcal{S}.$$

It can be seen that the closed-loop system in (6.10) is a delay-free fuzzy system. Next, we give the sufficient condition for system (6.10).

Theorem 6.1. *The fuzzy system (6.10) is globally asymptotically stable, if there exist matrices $P_{ij} > 0$, F_{ij} and G_{ij} with appropriate dimensions such that the following LMIs are satisfied:*

$$\Phi_{pqlii} < 0, \quad i, p, q \in \mathcal{N}, \ l \in \mathcal{S}, \quad (6.11)$$

$$\Phi_{pqlij} + \Phi_{pqlji} < 0, \quad i < j, i, j, p, q \in \mathcal{N}, \ l \in \mathcal{S}, \quad (6.12)$$

where

$$\Phi_{pqlij} \overset{\Delta}{=} \begin{bmatrix} -P_{pq} & F_{pq}^T \bar{A}_{ijl} & P_{pq} - F_{pq}^T \\ * & \Phi_{pqlij}^{(22)} & \bar{A}_{ijl}^T F_{pq} - G_{pq}^T \\ * & * & -F_{pq} - F_{pq}^T \end{bmatrix},$$

with $\Phi_{pqlij}^{(22)} = \bar{A}_{ijl}^T G_{pq} + G_{pq}^T \bar{A}_{ijl} - P_{ij}$.

Proof. To prove the theorem, we introduce an indicator function $\delta_{k,i}$ as

$$\delta_{k,i} = \begin{cases} 1, & i = d_k \\ 0, & i \neq d_k \end{cases}, \ i \in \mathcal{S}.$$

Therefore, the closed-loop system in (6.10) can be rewritten as

$$\bar{x}(k+1) = \bar{A}_{h,\delta} \bar{x}(k), \quad (6.13)$$

where $\bar{A}_{h,\delta} = \sum\limits_{i=1}^{r} \sum\limits_{j=1}^{r} \sum\limits_{l=d_m}^{d_M} h_i(\theta_k) h_j(\theta_k) \delta_{k,l} \bar{A}_{ijl}$ with

$$\bar{A}_{ijl} = \left[\begin{array}{c|ccc} A_i & 0_{n\times(l-1)n} & B_i K_{j,l} & 0_{n\times(d_M-l)n} \\ \hline I_{d_M n \times d_M n} & & 0_{d_M n \times n} \end{array} \right], \quad d_k \in \mathcal{S}.$$

On one hand, we take the fuzzy Lyapunov function as $V(k) = \bar{x}^T(k) \bar{P}_h \bar{x}(k)$, where $\bar{P}_h = \sum\limits_{i=1}^{r} \sum\limits_{j=1}^{r} h_i(\theta_k) h_j(\theta_k) P_{ij}$ with $P_{ij} > 0$.

Define $\Delta V(k) = V(k+1) - V(k)$. Along the solution of system (6.10), we have

$$\Delta V(k) = \bar{x}^T(k+1) \bar{P}_{h^+} \bar{x}(k+1) - \bar{x}^T(k) \bar{P}_h \bar{x}(k)$$
$$= \bar{x}^T(k) \left(\bar{A}_{h,\delta}^T \bar{P}_{h^+} \bar{A}_{h,\delta} - \bar{P}_h \right) \bar{x}(k).$$

On the other hand, by noticing that, at any time step k, $\sum\limits_{i=1}^{r} h_i(\theta_k) = 1$, $\sum\limits_{j=d_m}^{d_M} \delta_{k,j} = 1$ and $h_i(\theta(k)) \geq 0$, $\delta_{k,i} \geq 0$, then considering (6.11) and (6.12), we can conclude that the following matrix inequality holds

$$\left[\begin{array}{ccc} -P_{h^+} & F_{h^+}^T \bar{A}_{h,\delta} & P_{h^+} - F_{h^+}^T \\ * & \bar{A}_{h,\delta}^T G_{h^+} + G_{h^+}^T \bar{A}_{h,\delta} - P_h & \bar{A}_{h,\delta}^T F_{h^+} - G_{h^+}^T \\ * & * & -F_{h^+} - F_{h^+}^T \end{array} \right]$$

$$= \sum\limits_{p=1}^{r} \sum\limits_{q=1}^{r} \sum\limits_{i=1}^{r} \sum\limits_{j=1}^{r} \sum\limits_{l=d_m}^{d_M} h_p(\theta_{k+1}) h_q(\theta_{k+1}) h_i(\theta_k) h_j(\theta_k) \delta_{k,l} \Phi_{pqlij}$$

$$= \sum\limits_{p=1}^{r} \sum\limits_{q=1}^{r} \sum\limits_{l=d_m}^{d_M} h_p(\theta_{k+1}) h_q(\theta_{k+1}) \delta_{k,l} \left\{ \sum\limits_{i=1}^{r} h_i^2(\theta_k) \Phi_{pqlii} \right.$$

$$\left. + \sum\limits_{i=1}^{r-1} \sum\limits_{j=i+1}^{r} h_i(\theta_k) h_j(\theta_k) (\Phi_{pqlij} + \Phi_{pqlji}) \right\} < 0. \tag{6.14}$$

Pre- and post-multiplying (6.14) by $\left[\begin{array}{ccc} 0 & I & \bar{A}_{h,\delta}^T \\ I & 0 & 0 \end{array} \right]$ and its transpose yields that

$$\left[\begin{array}{cc} -P_h & \bar{A}_{h,\delta}^T P_{h^+} \\ P_{h^+} \bar{A}_{h,\delta} & -P_{h^+} \end{array} \right] < 0, \tag{6.15}$$

which is equivalent to $\bar{A}_{h,\delta}^T P_{h^+} \bar{A}_{h,\delta} - P_h < 0$. Then, there exists a positive scalar $\mu > 0$ such that $\bar{A}_{h,\delta}^T P_{h^+} \bar{A}_{h,\delta} - P_h + \mu I < 0$. Therefore, we have $\Delta V(k) \leq -\mu \|\bar{x}(k)\|^2$ for all nonzero $\bar{x}(k)$, and the asymptotic stability is established. $\square$

Based on the stability analysis results, in the following, we will further consider how to design stabilizing state feedback controllers.

Notice that $\bar{A}_{h,d_k}$ in (6.10) can be rewritten as

$$\bar{A}_{h,d_k} = \tilde{A}_h + \tilde{B}_h K_{h,d_k} \mathcal{I}_{d_k},$$

where

$$\tilde{A}_h = \left[\begin{array}{c|c} A_h & 0_{n \times d_M n} \\ \hline I_{d_M n} & 0_{d_M n \times n} \end{array} \right], \quad \tilde{B}_h = \left[\begin{array}{c} B_h \\ 0_{d_M n \times m} \end{array} \right], \quad \mathcal{I}_{d_k} = \left[\begin{array}{ccc} 0_{n \times d_k n} & I_n & 0_{n \times (d_M - d_k)n} \end{array} \right]$$

and $u^s(k|k - d_k) = K_{h,d_k} \mathcal{I}_{d_k} \bar{x}(k)$, we have

$$E \tilde{x}(k+1) = \mathcal{A}_{h,d_k} \tilde{x}(k) \tag{6.16}$$

where $\tilde{x}(k) = \left[\begin{array}{c} \bar{x}(k) \\ u^s(k|k - d_k) \end{array} \right]$, and $E = \left[\begin{array}{c|c} I_{(d_M+1)n} & 0_{(d_M+1)n \times m} \\ \hline 0_{m \times (d_M+1)n} & 0_m \end{array} \right]$, $\mathcal{A}_{h,d_k} = \left[\begin{array}{c|c} \tilde{A}_h & \tilde{B}_h \\ \hline K_{h,d_k} \mathcal{I}_{d_k} & -I_m \end{array} \right]$.

Next, based on singular fuzzy system (6.16), we give another stability condition for fuzzy system (6.10), which plays a key role in determining the state-based delay compensation controller.

Lemma 6.1. *Fuzzy system (6.10) is globally asymptotically stable if there exist matrices $P_{ij} > 0$, F_{ij}, G_{ij}, and Φ with appropriate dimensions such that the following LMIs are satisfied,*

$$\Sigma_{pqlii} < 0, \quad i, p, q \in \mathcal{N}, \ l \in \mathcal{S}, \tag{6.17}$$

$$\Sigma_{pqlij} + \Sigma_{pqlji} < 0, \quad i < j, i, j, p, q \in \mathcal{N}, \ l \in \mathcal{S}, \tag{6.18}$$

where

$$\Sigma_{pqlij} \overset{\Delta}{=} \left[\begin{array}{ccc} -P_{pq} & F_{pq}^T \mathcal{A}_{ijl} & P_{pq} - F_{pq}^T \\ * & \Sigma_{pqlij}^{(22)} & \mathcal{A}_{ijl}^T F_{pq} - G_{pq}^T \\ * & * & \Sigma_{pqlij}^{(33)} \end{array} \right],$$

with $E^\perp = \left[\begin{array}{cc} 0_{m \times (d_M+1)n} & I_m \end{array} \right]$, $\Sigma_{pqlij}^{(22)} = \mathcal{A}_{ijl}^T G_{pq} + G_{pq}^T \mathcal{A}_{ijl} - E^T P_{ij} E$, $\Sigma_{pqlij}^{(33)} = -F_{pq} - F_{pq}^T - (E^\perp)^T \Phi E^\perp$.

Proof. It follows from (6.17) and (6.18) that

$$\left[\begin{array}{ccc} -P_{h^+} & F_{h^+}^T \mathcal{A}_{h,\delta} & P_{h^+} - F_{h^+}^T \\ * & \mathcal{A}_{h,\delta}^T G_{h^+} + G_{h^+}^T \mathcal{A}_{h,\delta} - E^T P_h E & \mathcal{A}_{h,\delta}^T F_{h^+} - G_{h^+}^T \\ * & * & -F_{h^+} - F_{h^+}^T - (E^\perp)^T \Phi E^\perp \end{array} \right]$$

$$= \sum_{p=1}^{r} \sum_{q=1}^{r} \sum_{l=d_m}^{d_M} h_p(\theta_{k+1}) h_q(\theta_{k+1}) \delta_{k,l} \left\{ \sum_{i=1}^{r} h_i^2(\theta_k) \Sigma_{pqlii} \right.$$

$$\left. + \sum_{i=1}^{r-1} \sum_{j=i+1}^{r} h_i(\theta_k) h_j(\theta_k) (\Sigma_{pqlij} + \Sigma_{pqlji}) \right\} < 0.$$

Pre- and post-multiplying the above inequality by $\left[\begin{array}{ccc} 0 & I & \mathcal{A}_{h,\delta}^T \\ I & 0 & 0 \end{array} \right]$ and its transpose implies that $\mathcal{A}_{h,d_k}^T \left(P_{h^+} - (E^\perp)^T \Phi E^\perp \right) \mathcal{A}_{h,d_k} - E^T P_h E < 0$.

Note that

$$
\mathcal{A}_{h,d_k} = \left[\begin{array}{c|c} \tilde{A}_h + \tilde{B}_h K_{h,d_k} \mathcal{I}_{d_k} & \tilde{B}_h \\ \hline 0 & -I_m \end{array} \right] \left[\begin{array}{cc} I_{(d_M+1)n \times (d_M+1)n} & 0 \\ -K_{h,d_k} \mathcal{I}_{d_k} & I_m \end{array} \right]
$$

and denote $P_{h+} = \left[\begin{array}{cc} P_{h+}^{11} & (*) \\ (*) & (*) \end{array} \right]$, $P_h = \left[\begin{array}{cc} P_h^{11} & (*) \\ (*) & (*) \end{array} \right]$, with $P_{h+}^{11} > 0$ and $P_h^{11} > 0$, where $(*)$ denotes the matrix items are not used later. Then, we have

$$
\mathcal{A}_{h,d_k}^T \left(P_{h+} - (E^\perp)^T \Phi E^\perp \right) \mathcal{A}_{h,d_k} - E^T P_h E
$$
$$
= \left[\begin{array}{cc} I_{(d_M+1)n} & 0 \\ -\mathcal{I}_{d_k} & I_m \end{array} \right]^T \left[\begin{array}{cc} \bar{A}_{h,d_k}^T P_{h+}^{11} \bar{A}_{h,d_k} - P_h^{11} & (*) \\ (*) & (*) \end{array} \right] \left[\begin{array}{cc} I_{(d_M+1)n} & 0 \\ -\mathcal{I}_{d_k} & I_m \end{array} \right] < 0. \quad (6.19)
$$

It follows from (6.19) that $\bar{A}_{h,d_k}^T P_{h+}^{11} \bar{A}_{h,d_k} - P_h^{11} < 0$, which implies that system (6.10) is globally asymptotically stable according to Theorem 6.1, and this completes the proof. □

Now, we are in a position to design the state-based delay compensation controller (6.3).

Theorem 6.2. *Fuzzy system (6.10) is globally asymptotically stable if there exist matrices $P_{ij} > 0$, F_{ij}, G_{ij}, L_i, J and Φ with appropriate dimensions such that the following LMIs are satisfied,*

$$
\Theta_{pqlii} < 0, \quad i, p, q \in \mathcal{N}, \ l \in \mathcal{S}, \quad (6.20)
$$
$$
\Theta_{pqlij} + \Theta_{pqlji} < 0, \quad i < j, p, q \in \mathcal{N}, \ l \in \mathcal{S}, \quad (6.21)
$$

where

$$
\Theta_{pqlij} \triangleq \left[\begin{array}{ccc} -P_{pq} & \Pi_{pqlij} & P_{pq} - F_{pq}^T \\ * & \Theta_{pqlij}^{(22)} & \Pi_{pqlij}^T - G_{pq}^T \\ * & * & \Theta_{pqlij}^{(33)} \end{array} \right],
$$

$$
\Theta_{pqlij}^{(22)} = \Lambda_{pqijl}^T + \Lambda_{pqijl} - E^T P_{ij} E, \quad \Theta_{pqlij}^{(33)} = -F_{pq} - F_{pq}^T - (E^\perp)^T \Phi E^\perp,
$$

$$
\Pi_{pqlij} = \left[\begin{array}{cc} F_{1pq}^T \tilde{A}_i + \mathcal{J}^T L_{jl} \mathcal{I}_l & F_{1pq}^T \tilde{B}_i - \mathcal{J}^T J^T \\ F_{2pq}^T \tilde{A}_i + L_{jl} \mathcal{I}_l & F_{2pq}^T \tilde{B}_i - J^T \end{array} \right],
$$

$$
\Lambda_{pqlij} = \left[\begin{array}{cc} G_{1pq}^T \tilde{A}_i + \mathcal{J}^T L_{jl} \mathcal{I}_l & G_{1pq}^T \tilde{B}_i - \mathcal{J}^T J^T \\ G_{2pq}^T \tilde{A}_i + L_{jl} \mathcal{I}_l & G_{2pq}^T \tilde{B}_i - J^T \end{array} \right],
$$

and $\mathcal{J} = [I_m, 0_{m \times ((d_M+1)n-m)}]$ and $F_{pq} = \left[\begin{array}{cc} F_{1pq} & F_{2pq} \\ JJ & J \end{array} \right]$, $G_{pq} = \left[\begin{array}{cc} G_{1pq} & G_{2pq} \\ JJ & J \end{array} \right]$.
Furthermore, if the aforementioned conditions are feasible, the state-based delay compensation controller gain can be given by $K_{il} = J^{-1} L_{il}$.

Proof. By noticing the special structure of F_{pq} and G_{pq}, and substituting them as well as $L_{il} = JK_{il}$ to (6.17) and (6.18), one can obtain (6.20) and (6.21) immediately, and the proof is completed. □

Next, we will consider the output feedback case. To this end, we first rewrite system (6.8) into the following form:

$$x(k + 1) = A_h x(k) + B_h H_{h,d_k} y(k - d_k) \qquad (6.22)$$

where A_h and B_h are given in (6.9), and $H_{h,d_k} = \sum_{i=1}^{r} h_i(\theta_k) H_{id_k}$.

To analyze the stability of system (6.22), we introduce the following augmented state vector $\hat{x}(k) = \begin{bmatrix} x^T(k) & y^T(k-1) & \cdots & y^T(k-d_M) \end{bmatrix}^T$. Thus, when the time-varying network communication delay d_k takes value in $\mathcal{S}$, system (6.22) can be augmented as

$$\hat{x}(k + 1) = \hat{A}_{h,d_k} \hat{x}(k) \qquad (6.23)$$

where

$$\hat{A}_{h,d_k} = \left[\begin{array}{c|cc} A_h & 0_{n\times(d_k-1)\nu} \quad B_h H_{h,d_k} \quad 0_{n\times(d_M-d_k)\nu} \\ \hline C_h & 0_{\nu\times d_M \nu} \\ \hline 0_{(d_M-1)\nu\times n} & I_{(d_M-1)\nu} \quad 0_{(d_M-1)\nu\times\nu} \end{array} \right].$$

Theorem 6.3. *The fuzzy system (6.23) is globally asymptotically stable, if there exist matrices $P_{ij} > 0$, F_{ij}, and G_{ij} with appropriate dimensions such that the following LMIs are satisfied:*

$$\Psi_{pqlii} < 0, \quad i, p, q \in \mathcal{N}, \ l \in \mathcal{S}, \qquad (6.24)$$

$$\Psi_{pqlij} + \Psi_{pqlji} < 0, \quad i < j, i, j, p, q \in \mathcal{N}, \ l \in \mathcal{S}, \qquad (6.25)$$

where

$$\Psi_{pqlij} \triangleq \begin{bmatrix} -P_{pq} & F_{pq}^T \hat{A}_{ijl} & P_{pq} - F_{pq}^T \\ * & \Psi_{pqlij}^{(22)} & \hat{A}_{ijl}^T F_{pq} - G_{pq}^T \\ * & * & -F_{pq} - F_{pq}^T \end{bmatrix},$$

with $\Psi_{pqlij}^{(22)} = \hat{A}_{ijl}^T G_{pq} + G_{pq}^T \hat{A}_{ijl} - P_{ij}$.

In the following, we will consider how to design output-based delay compensation controller (6.5). Noticing that $\hat{A}_{h,d_k}$ in (6.23) can be rewritten as

$$\hat{A}_{h,d_k} = \check{A}_h + \check{B}_h H_{h,d_k} \check{I}_{d_k},$$

where

$$\check{A}_h = \left[\begin{array}{c|cc} A_h & 0_{n\times d_M\nu} \\ \hline C_h & 0_{\nu\times d_M\nu} \\ \hline 0_{(d_M-1)\nu\times n} & I_{(d_M-1)\nu} \quad 0_{(d_M-1)\nu\times\nu} \end{array} \right], \check{B}_h = \begin{bmatrix} B_h \\ 0_{d_M\nu\times m} \end{bmatrix},$$

$$\check{I}_{d_k} = \begin{bmatrix} 0_{\nu\times n} & 0_{\nu\times(d_k-1)\nu} & I_\nu & 0_{\nu\times(d_M-d_k)\nu} \end{bmatrix},$$

and $u^o(k|k - d_k) = H_{h,d_k}\breve{\mathcal{I}}_{d_k}\hat{x}(k)$, we have

$$\breve{E}\breve{x}(k+1) = \breve{\mathcal{A}}_{h,d_k}\breve{x}(k) \tag{6.26}$$

where $\breve{x}(k) = \begin{bmatrix} \hat{x}(k) \\ u^o(k|k - d_k) \end{bmatrix}$, and $\breve{E} = \begin{bmatrix} I_{n+d_M\nu} & 0_{(n+d_M\nu)\times m} \\ \hline 0_{m\times(n+d_M\nu)} & 0_m \end{bmatrix}$, $\breve{\mathcal{A}}_{h,d_k} = \begin{bmatrix} \breve{A}_h & \breve{B}_h \\ \hline H_{h,d_k}\breve{\mathcal{I}}_{d_k} & -I_m \end{bmatrix}$.

Next, based on singular fuzzy system (6.26), another sufficient condition is proposed for the stability of fuzzy system (6.23), which will be useful for determining the output-based delay compensation controller. Similar to Lemma 6.1, we have the following lemma.

Lemma 6.2. *Fuzzy system (6.23) is globally asymptotically stable if there exist matrices $P_{ij} > 0$, F_{ij}, G_{ij}, and Φ with appropriate dimensions such that the following LMIs are satisfied,*

$$\breve{\Sigma}_{pqlii} < 0, \quad i, p, q \in \mathcal{N}, \ l \in \mathcal{S}, \tag{6.27}$$

$$\breve{\Sigma}_{pqlij} + \breve{\Sigma}_{pqlji} < 0, \quad i < j, i, j, p, q \in \mathcal{N}, \ l \in \mathcal{S}, \tag{6.28}$$

where

$$\breve{\Sigma}_{pqlij} \triangleq \begin{bmatrix} -P_{pq} & F_{pq}^T\breve{\mathcal{A}}_{ijl} & P_{pq} - F_{pq}^T \\ * & \breve{\Sigma}_{pqlij}^{(22)} & \breve{\mathcal{A}}_{ijl}^T F_{pq} - G_{pq}^T \\ * & * & \breve{\Sigma}_{pqlij}^{(33)} \end{bmatrix},$$

with $\breve{E}^\perp = \begin{bmatrix} 0_{m\times(n+d_M\nu)} & I_m \end{bmatrix}$, and $\breve{\Sigma}_{pqlij}^{(22)} = \breve{\mathcal{A}}_{ijl}^T G_{pq} + G_{pq}^T\breve{\mathcal{A}}_{ijl} - \breve{E}^T P_{ij}\breve{E}$, $\breve{\Sigma}_{pqlij}^{(33)} = -F_{pq} - F_{pq}^T - (\breve{E}^\perp)^T\Phi\breve{E}^\perp$.

Now, we are ready to design the output-based delay compensation controller (6.5).

Theorem 6.4. *Fuzzy system (6.23) is globally asymptotically stable if there exist matrices $P_{ij} > 0$, F_{ij}, G_{ij}, L_i, J, and Φ with appropriate dimensions such that the following LMIs are satisfied,*

$$\breve{\Theta}_{pqlii} < 0, \quad i, p, q \in \mathcal{N}, \ l \in \mathcal{S}, \tag{6.29}$$

$$\breve{\Theta}_{pqlij} + \breve{\Theta}_{pqlji} < 0, \quad i, j, p, q \in \mathcal{N}, \ l \in \mathcal{S}, \tag{6.30}$$

where

$$\breve{\Theta}_{pqlij} \triangleq \begin{bmatrix} -P_{pq} & \breve{\Pi}_{pqijl} & P_{pq} - F_{pq}^T \\ * & \breve{\Theta}_{pqlij}^{(22)} & \breve{\Pi}_{pqijl}^T - G_{pq}^T \\ * & * & \breve{\Theta}_{pqlij}^{(33)} \end{bmatrix},$$

$$\breve{\Theta}_{pqlij}^{(22)} = \breve{\Lambda}_{pqijl}^T + \breve{\Lambda}_{pqijl} - \breve{E}^T P_{ij}\breve{E}, \quad \breve{\Theta}_{pqlij}^{(33)} = -F_{pq} - F_{pq}^T - (\breve{E}^\perp)^T\Phi\breve{E}^\perp,$$

$$\breve{\Pi}_{pqijl} = \begin{bmatrix} F_{1pq}^T\breve{A}_i + \breve{J}^T L_{jl}\mathcal{I}_l & F_{1pq}^T\breve{B}_i - \breve{J}^T J^T \\ F_{2pq}^T\breve{A}_i + L_{jl}\mathcal{I}_l & F_{2pq}^T\breve{B}_i - J^T \end{bmatrix},$$

$$\breve{\Lambda}_{pqijl} = \begin{bmatrix} G_{1pq}^T\breve{A}_i + \breve{J}^T L_{jl}\mathcal{I}_l & G_{1pq}^T\breve{B}_i - \breve{J}^T J^T \\ G_{2pq}^T\breve{A}_i + L_{jl}\mathcal{I}_l & G_{2pq}^T\breve{B}_i - J^T \end{bmatrix},$$

and $\check{J} = [I_m, 0_{m \times (n+d_M \nu - m)}]$ and $F_{pq} = \begin{bmatrix} F_{1pq} & F_{2pq} \\ J\check{J} & J \end{bmatrix}$, $G_{pq} = \begin{bmatrix} G_{1pq} & G_{2pq} \\ J\check{J} & J \end{bmatrix}$.

Furthermore, if the aforementioned conditions are feasible, the output-based delay compensation controller gain can be given by $H_{il} = J^{-1}L_{il}$.

Proof. By using Lemma 6.2, it is similar to the proof of Theorem 6.2. The details are omitted here. □

6.5 NUMERICAL SIMULATION

To illustrate the theoretical results developed in this chapter, we consider the inverted pendulum controlled by a DC motor via a gear train shown in Figure 6.2, whose discrete-time model is:

$$x(k+1) = \sum_{i=1}^{2} h_i(A_i x(k) + B_{ui} u(k))$$

where

$$A_1 = \begin{bmatrix} 1.002 & 0.02 & 0.02 \\ 0.196 & 1.0001 & 0.0181 \\ -0.0184 & -0.1813 & 0.8170 \end{bmatrix}, A_2 = \begin{bmatrix} 1 & 0.02 & 0.0002 \\ 0 & 0.9981 & 0.0181 \\ 0 & -0.1811 & 0.8170 \end{bmatrix},$$

$$B_{u1} = B_{u2} = \begin{bmatrix} 0 \\ 0.0019 \\ 0.1811 \end{bmatrix}.$$

To show the effectiveness of the obtained results, we assume the membership function at the plant and controller side to be $h_1(x_{1,k}) = 1 - 0.25x_{1,k}^2$, and $h_2(x_{1,k}) = 1 - h_1(x_{1,k})$.

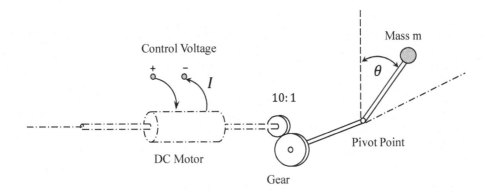

Figure 6.2 Inverted pendulum controlled by a DC motor

Now, we assume the inverted pendulum is controlled over network, and lower and upper bound of the networked communication delay d_k is 1 and 3, respectively.

According to Theorem 6.2, we can get the fuzzy delay compensation controller gains are

$$K_{11} = [\begin{array}{ccc} -2.5771 & -0.2831 & 0.0412 \end{array}], \quad K_{12} = [\begin{array}{ccc} -2.5880 & -0.3216 & 0.0234 \end{array}],$$
$$K_{13} = [\begin{array}{ccc} -2.8233 & -0.3818 & -0.0465 \end{array}], \quad K_{21} = [\begin{array}{ccc} -1.7863 & -0.0539 & 0.3138 \end{array}],$$
$$K_{22} = [\begin{array}{ccc} -1.6190 & -0.0113 & 0.3626 \end{array}], \quad K_{23} = [\begin{array}{ccc} -1.7762 & -0.0418 & 0.3150 \end{array}].$$

The initial condition is chosen as $x_0 = \begin{bmatrix} 0.8 & -0.4 & -0.8 \end{bmatrix}^T$, by applying the fuzzy delay compensation controller designed in Theorem 6.2, the simulation results are shown in Figure 6.3(a–d).

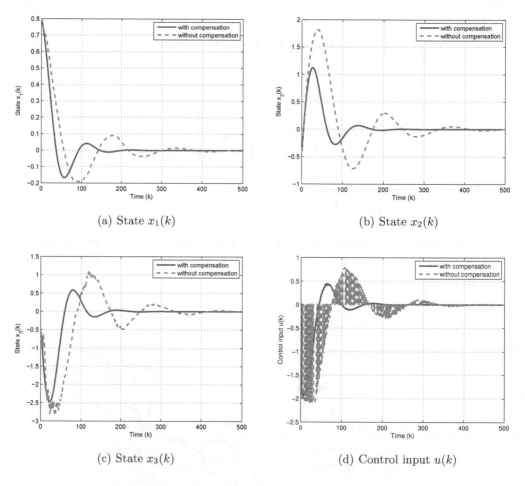

(a) State $x_1(k)$ (b) State $x_2(k)$

(c) State $x_3(k)$ (d) Control input $u(k)$

Figure 6.3 States and control input of inverted pendulum system (state-feedback case)

From those simulation results, it is clear that fuzzy delay compensation scheme proposed in this chapter can compensate for the network communication delay actively.

In the following, to show the effectiveness of output feedback network delay compensation controller design scheme, we add the measurement equation,

$$y(k) = \begin{bmatrix} 1 & 0 & 0 \end{bmatrix} x(k).$$

According to Theorem 6.4, we can obtain the following controller gains,

$$K_{11} = -3.4990, \ K_{12} = -3.4406, \ K_{13} = -3.5047,$$
$$K_{21} = -3.1287, \ K_{22} = -3.1472, \ K_{23} = -3.1366.$$

By using the output fuzzy delay compensation controller, the simulation results are shown in Figure 6.4(a–d), which also show the effectiveness of the controller design.

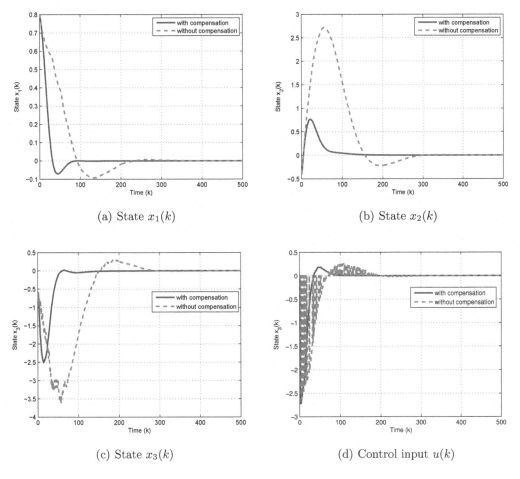

(a) State $x_1(k)$

(b) State $x_2(k)$

(c) State $x_3(k)$

(d) Control input $u(k)$

Figure 6.4 States and control input of inverted pendulum system (output-feedback case)

6.6 SUMMARY

In this chapter, the network delay compensation control problem for nonlinear plant in CPSs has been investigated. By using packet-based technique, both state feedback and output feedback fuzzy delay compensation controllers have been developed to actively compensate the network communication delay. The stability of networked closed-loop system has been analyzed, and the fuzzy delay compensation controllers has been designed by using singular fuzzy system approach. Numerical example has been given to illustrate the developed theoretical results.

CHAPTER 7

Networked Output Tracking Control of CPSs with Delay Compensation

7.1 INTRODUCTION

In the networked delay compensation control approaches proposed in Chapter 4, Chapter 5, and Chapter 6, the CIPs are computed based on the delayed state/output measurements [228, 223, 207, 204], which indicates the existing networked delay compensation control methods are only effective in compensating for the delay in the controller-to-actuator channel in CPSs, and how to compensate the sensor-to-controller channel delay is also important and interesting. Moreover, the output tracking control has wide applications in dynamic processes in industry, economics, and biology. The main objective of output tracking control is to design suitable controller to make the output of the plant track a given reference signal as close as possible. In recent years, some works have been concerned with the output tracking problem of CPSs [195, 39, 162, 193]. In [195], the delay-dependent tracking controller is designed for CPSs with Markov delays. In [39, 162], the H_∞ output tracking problems are investigated for CPSs with both the network communication delay and data dropout. The formation tracking control problem is studied for multiple spacecrafts with network communication delay and bounded external disturbance. In [193], the network-based output tracking control for a continuous-time T-S fuzzy system is considered, and the event-triggered communication scheme is proposed to decide whether or not the sampled-data should be transmitted through a communication network.

In this chapter, the problem of networked output tracking control is investigated by considering the delay compensations in both the sensor-to-controller and controller-to-actuator channels in CPSs. The delayed output measurements are treated as a special output disturbance, and the sensor-to-controller channel delay is compensated with the aid of an extended functional state observer (EFSO). For the delay in the controller-to-actuator channel, the buffer and packet-based delay compensation approaches are presented, respectively. Then, the stability analysis is performed for the networked closed-loop systems. Finally, a servo motor control system is used to demonstrate the effectiveness of the proposed control approaches.

7.2 SYSTEM DESCRIPTION

The CPSs structure considered in this chapter is shown in Figure 7.1, where d_k^{sc} and

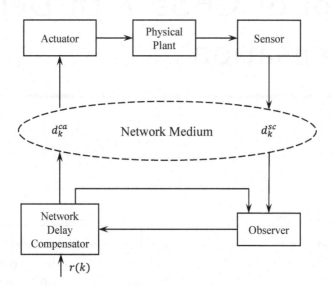

Figure 7.1 Block diagram of networked tracking control systems

d_k^{ca} are time-varying network communication delays in the sensor-to-controller and controller-to-actuator network channels, respectively. The plant considered in this chapter can be described by the following state space model:

$$x(k+1) = Ax(k) + Bu(k)$$
$$y(k) = Cx(k) \tag{7.1}$$

where $x(k) \in \mathbb{R}^n$ is the state vector, $u(k) \in \mathbb{R}^m$ is the control input vector, $y(k) \in \mathbb{R}^q$ is the system output vector, $r(k)$ is a bounded reference input, and has constant steady-state value. A, B, and C are known constant system matrices with appropriate dimensions. In addition, the following assumptions are made for system (7.1).

Assumption 7.1. *The pair (A, B) is stabilizable, the pair (A, C) is detectable, and*
$\begin{bmatrix} A - I_n & B \\ C & 0 \end{bmatrix}$ *is of full column rank.*

Assumption 7.2. *The time-varying network communication delays d_k^{sc} in the sensor-to-controller channel and d_k^{ca} in the controller-to-actuator channel are bounded by $\bar{d}_{sc}$ and $\bar{d}_{ca}$, respectively, where $\bar{d}_{sc}$ and $\bar{d}_{ca}$ are two positive integers.*

Assumption 7.3. *The observer and actuator possess the logical choice capability to guarantee that only the latest data will be used.*

Remark 7.1. *In this chapter, only network communication delay is considered in both sensor-to-controller and controller-to-actuator channels. It should be pointed out that, if the data dropout happens in the sensor-to-controller or controller-to-actuator channel, under Assumption 7.3, the data dropout can be viewed as a special network communication delay and merged into d_k^{sc} or d_k^{ca}.*

The objective of this chapter is to propose the delay compensation control schemes to actively compensate the delay occurred in both sensor-to-controller and controller-to-actuator channels, and to design the controller such that the output $y(k)$ tracks the given reference signal $r(k)$.

7.3 DELAY COMPENSATION IN SENSOR-TO-CONTROLLER NETWORK CHANNEL

In this section, we consider the delay compensation control problem in the sensor-to-controller channel. As shown in Figure 7.1, the observer is located at the controller side, and there exists network communication delay d_k^{sc} in the sensor-to-controller channel, thus, at instant k, the corresponding output signal received by the observer is $y_{sc}(k) \triangleq y(k - d_k^{sc})$ rather than $y(k)$. If the system state is not available for the system, we need to introduce an observer to estimate the state. The common and traditional observer is given as

$$\hat{x}(k + 1) = A\hat{x}(k) + Bu(k) + L(y_{sc}(k) - C\hat{x}(k)). \tag{7.2}$$

Note that the system output at the observer side can be rewritten in the following form:

$$\begin{aligned} y_{sc}(k) = y(k - d_k^{sc}) &= Cx(k - d_k^{sc}) \\ &= Cx(k) + x_f(k) \end{aligned}$$

where $x_f(k) = Cx(k - d_k^{sc}) - Cx(k)$ can be viewed as the "disturbance" induced by the network in the sensor-to-controller channel. If the observer is chosen as (7.2), the estimation error $e_x(k) \triangleq x(k) - \hat{x}(k)$ can be determined as

$$\begin{aligned} e_x(k + 1) &= x(k + 1) - \hat{x}(k + 1) \\ &= Ax(k) + Bu(k) - (A\hat{x}(k) + Bu(k) + L(y_{sc}(k) - C\hat{x}(k))) \\ &= Ax(k) - (A\hat{x}(k) + L(Cx(k) + x_f(k) - C\hat{x}(k))) \\ &= (A - LC)e_x(k) - Lx_f(k). \tag{7.3} \end{aligned}$$

Obviously, the estimation error will be affected unavoidably by the "disturbance" $x_f(k)$ and the observer gain L, and system performance will be degraded inevitably by using the traditional observer (7.2) for system (7.1). Motivated by the state observer technique proposed in [31, 57, 58, 149], in the following, the EFSO will be designed for system (7.1) to estimate the state $x(k)$ and the "disturbance" $x_f(k)$ simultaneously.

Now, for the simplicity of presentation, we introduce the following notations:

$$\bar{x}(k) = \begin{bmatrix} x(k) \\ x_f(k) \end{bmatrix} \quad \bar{E} = \begin{bmatrix} I_n & 0_{n \times q} \end{bmatrix},$$

$$\bar{A} = \begin{bmatrix} A & 0_{n \times q} \end{bmatrix}, \quad \bar{C} = \begin{bmatrix} C & I_q \end{bmatrix}.$$

As a result, system (7.1) at the observer side can be described in the following extended form:

$$\bar{E}\bar{x}(k+1) = \bar{A}\bar{x}(k) + Bu(k),$$
$$y_{sc}(k) = \bar{C}\bar{x}(k). \tag{7.4}$$

Then, the state and "disturbance" estimation problem can be converted into the problem of designing an observer for the extended system (7.4). Now, consider the following EFSO,

$$\eta(k+1) = F\hat{\bar{x}}(k) + Gy_{sc}(k) + Hu(k)$$
$$\hat{\bar{x}}(k) = \eta(k) + Qy_{sc}(k), \tag{7.5}$$

where $\hat{\bar{x}}(k)$ denotes the estimate of $\bar{x}(k)$, $\eta(k)$ is an auxiliary variable, matrices F, G, H, Q are observer parameters to be determined such that $\hat{\bar{x}}(k)$ converges asymptotically to $\bar{x}(k)$.

Note that

$$\text{rank} \begin{bmatrix} \bar{E} \\ \bar{C} \end{bmatrix} = \text{rank} \begin{bmatrix} I_n & 0_{n \times q} \\ C & I_q \end{bmatrix} = n + q,$$

that is, the matrix $\begin{bmatrix} \bar{E} \\ \bar{C} \end{bmatrix}$ is of full column rank, which implies that $\begin{bmatrix} \bar{E} \\ \bar{C} \end{bmatrix}^{-1}$ exists. Denote

$$M = \begin{bmatrix} \bar{E} \\ \bar{C} \end{bmatrix}^{-1} \begin{bmatrix} I_n \\ 0_{q \times n} \end{bmatrix} = \begin{bmatrix} I_n \\ -C \end{bmatrix},$$

$$N = \begin{bmatrix} \bar{E} \\ \bar{C} \end{bmatrix}^{-1} \begin{bmatrix} 0_{n \times q} \\ I_q \end{bmatrix} = \begin{bmatrix} 0_{n \times q} \\ I_q \end{bmatrix}.$$

Then, it is easy to obtain that

$$\begin{bmatrix} M & N \end{bmatrix} = \begin{bmatrix} \bar{E} \\ \bar{C} \end{bmatrix}^{-1}, \tag{7.6}$$

$$M\bar{E} + N\bar{C} = I_{n+q}. \tag{7.7}$$

Theorem 7.1. *For system (7.4), the estimation error $e_{\bar{x}}(k) = \bar{x}(k) - \hat{\bar{x}}(k)$ converges asymptotically to zero, if there exists a matrix $\bar{L}$ such that $M\bar{A} - \bar{L}\bar{C}$ is Schur. Furthermore, parameter matrices F, G, H, and Q of the EFSO (7.5) can be determined as*

$$F = M\bar{A} - \bar{L}\bar{C}, \quad G = \bar{L} + (M\bar{A} - \bar{L}\bar{C})N, \quad H = MB, \quad Q = N. \tag{7.8}$$

Proof. Multiplying M at both sides of (7.4) yields that:

$$M\bar{E}\bar{x}(k+1) = M\bar{A}\bar{x}(k) + MBu(k), \tag{7.9}$$

and adding $Ny_{sc}(k+1)$ to both sides of (7.9) and applying equation (7.7) yields that:

$$\bar{x}(k+1) = M\bar{A}\bar{x}(k) + MBu(k) + Ny_{sc}(k+1). \tag{7.10}$$

Denote $\vartheta(k) = MBu(k) + Ny_{sc}(k+1)$ as the input signal of system (7.10), then system (7.10) can be reconstructed as

$$\bar{x}(k+1) = M\bar{A}\bar{x}(k) + \vartheta(k)$$
$$y_{sc}(k) = \bar{C}\bar{x}(k).$$

Thus, we can design the following observer,

$$\begin{aligned}\hat{\bar{x}}(k+1) &= M\bar{A}\hat{\bar{x}}(k) + \vartheta(k) + \bar{L}(y_{sc}(k) - \bar{C}\hat{\bar{x}}(k)) \\ &= M\bar{A}\hat{\bar{x}}(k) + MBu(k) + Ny_{sc}(k+1) \\ &\quad + \bar{L}(y_{sc}(k) - \bar{C}\hat{\bar{x}}(k)).\end{aligned} \tag{7.11}$$

To eliminate the term $y_{sc}(k+1)$ from observer (7.11), we introduce an auxiliary variable $\eta(k) = \hat{\bar{x}}(k) - Ny_{sc}(k)$, and the observer (7.11) turns into

$$\begin{aligned}\eta(k+1) &= M\bar{A}\hat{\bar{x}}(k) + MBu(k) + \bar{L}(y_{sc}(k) - \bar{C}\hat{\bar{x}}(k)) \\ &= (M\bar{A} - \bar{L}\bar{C})\hat{\bar{x}}(k) + MBu(k) + \bar{L}y_{sc}(k) \\ &= (M\bar{A} - \bar{L}\bar{C})\eta(k) + MBu(k) \\ &\quad + (\bar{L} + (M\bar{A} - \bar{L}\bar{C})N)y_{sc}(k).\end{aligned} \tag{7.12}$$

Subtracting (7.11) from (7.10) yields the following estimation error equation

$$e_{\bar{x}}(k+1) = (M\bar{A} - \bar{L}\bar{C})e_{\bar{x}}(k) \tag{7.13}$$

which implies that the estimation error converges asymptotically to zero if $M\bar{A} - \bar{L}\bar{C}$ is Schur. Finally, comparing (7.12) with $\hat{\bar{x}}(k) = \eta(k) + Ny_{sc}(k)$ and observer (7.5) yields the observer parameters (7.8). Thus the proof is completed. □

In the following, we will show the existence of matrix $\bar{L}$ such that $M\bar{A} - \bar{L}\bar{C}$ is Schur. Let's recall the following useful lemma.

Lemma 7.1. *(Popov-Belevitch-Hautus Rank Test for Detectability). The pair (A, C) is detectable if and only if*

$$rank \begin{bmatrix} \lambda I_n - A \\ C \end{bmatrix} = n, \ \lambda \in \mathbb{C}, \ |\lambda| \geq 1.$$

Theorem 7.2. *There exists the observer gain $\bar{L}$ for the observer (7.12) such that $M\bar{A} - \bar{L}\bar{C}$ is Schur if and only if matrix A is Schur.*

Proof. It is easy to obtain that

$$M\bar{A} = \begin{bmatrix} I_n \\ -C \end{bmatrix} \begin{bmatrix} A & 0_{n\times q} \end{bmatrix} = \begin{bmatrix} A & 0_{n\times q} \\ -CA & 0_q \end{bmatrix}$$

and note that, for $\forall |\lambda| \geq 1$,

$$\text{rank}\left(\begin{bmatrix} \lambda I_{n+q} - M\bar{A} \\ \bar{C} \end{bmatrix}\right)$$

$$= \text{rank}\left(\begin{bmatrix} \lambda I_{n+q} - \begin{bmatrix} A & 0_{n\times q} \\ -CA & 0_q \end{bmatrix} \\ \bar{C} \end{bmatrix}\right)$$

$$= \text{rank}\left(\begin{bmatrix} I_n & 0_{n\times q} & 0_{n\times q} \\ -C & I_q & \lambda I_q \\ 0_{q\times n} & 0_q & I_q \end{bmatrix} \begin{bmatrix} \lambda I_n - A & 0_{n\times q} \\ 0_{q\times n} & 0_q \\ C & I_q \end{bmatrix}\right)$$

$$= \text{rank}\left(\begin{bmatrix} \lambda I_n - A & 0_{n\times q} \\ C & I_q \end{bmatrix}\right)$$

$$= \text{rank}(\lambda I_n - A) + q$$

which together with Lemma 7.1 implies that the pair $(M\bar{A}, \bar{C})$ is detectable if and only of A is Schur. Thus, if A is Schur, there exists a matrix $\bar{L}$ such that $M\bar{A} - \bar{L}\bar{C}$ is Schur. ◻

With the extend functional observer (7.5), the system state $x(k)$ and the "disturbance" $x_f(k)$ can be estimated asymptotically, and the estimates can be determined as $\hat{x}(k) = \begin{bmatrix} I_n & 0_{n\times q} \end{bmatrix} \hat{\bar{x}}(k)$, $\hat{x}_f(k) = \begin{bmatrix} 0_{q\times n} & I_q \end{bmatrix} \hat{\bar{x}}(k)$.

Remark 7.2. *The objective of introducing the EFSO (7.5) is to improve the estimation precision of system states. According to the estimation error equation (7.13), we can conclude that if $\bar{L}$ is selected such that $M\bar{A} - \bar{L}\bar{C}$ is Schur, the estimation error $e_{\bar{x}}(k)$ will converge asymptotically to zero. However, it can be observed from the estimation error equation (7.3) that, even if L is selected such that $A - LC$ is Schur, the estimation error $e_x(k)$ by using the traditional observer (7.2) will be affected by the disturbance $x_f(k)$. As a result, the proposed EFSO (7.5) is more suitable to obtain the state estimates.*

Remark 7.3. *According to Theorem 7.2, the EFSO (7.5) with parameters (7.8) is only suitable for the case that system matrix A in (7.1) is Schur. However, if A is not Schur and the pair (A, C) is detectable, we can design the EFSO in a similar way. Reformulating extended system (7.4) as*

$$\bar{\mathcal{E}}\bar{x}(k+1) = \bar{\mathcal{A}}\bar{x}(k) + \mathcal{B}u(k) + \mathcal{F}x_f(k)$$
$$y_{sc}(k) = \bar{\mathcal{C}}\bar{x}(k)$$

where $\bar{x}(k)$, $x_f(k)$ and $\bar{C}$ are the same as the ones defined in (7.4), and

$$\mathcal{E} = \begin{bmatrix} I_n & 0_{n \times q} \\ 0_{q \times n} & 0_q \end{bmatrix}, \quad \bar{\mathcal{A}} = \begin{bmatrix} A & 0_{n \times q} \\ 0_{q \times n} & -I_q \end{bmatrix}, \quad \mathcal{B} = \begin{bmatrix} B \\ 0_{q \times m} \end{bmatrix}, \quad \mathcal{F} = \begin{bmatrix} 0_{n \times q} \\ I_q \end{bmatrix}.$$

The matrices $\mathcal{M}$ and $\mathcal{N}$ satisfying the equation $\mathcal{M}\mathcal{E} + \mathcal{N}\bar{C} = I_{n+q}$ are selected as:

$$\mathcal{M} = \begin{bmatrix} I_n & 0_{n \times q} \\ -C & \epsilon_0 I_q \end{bmatrix}, \quad \mathcal{N} = \begin{bmatrix} 0_{n \times q} \\ I_q \end{bmatrix},$$

where ϵ_0 is a small enough positive scalar. Then, the following EFSO can be designed as

$$\eta(k+1) = (\mathcal{M}\bar{\mathcal{A}} - \bar{\mathcal{L}}\bar{C})\eta(k) + \mathcal{M}\mathcal{B}u(k)$$
$$+ (\bar{\mathcal{L}} + (\mathcal{M}\bar{\mathcal{A}} - \bar{\mathcal{L}}\bar{C})\mathcal{N})y_{sc}(k)$$
$$\hat{\bar{x}}(k) = \eta(k) + \mathcal{N}y_{sc}(k),$$

where the gain matrix $\bar{\mathcal{L}}$ is selected such that $\mathcal{M}\bar{\mathcal{A}} - \bar{\mathcal{L}}\bar{C}$ is Schur. Notice that, for $\forall |\lambda| > 1$, we have

$$rank \begin{bmatrix} \lambda I_{n+q} - \mathcal{M}\bar{\mathcal{A}} \\ \bar{C} \end{bmatrix} = rank \begin{bmatrix} \lambda I_n - A & 0_{n \times q} \\ CA & (\lambda + \epsilon_0)I_q \\ C & I_q \end{bmatrix}$$

$$= rank \begin{bmatrix} I_n & 0_{n \times q} & 0_{n \times q} \\ -C & \lambda I_q & (\lambda + \epsilon_0)I_q \\ 0 & I_q & I_q \end{bmatrix} \begin{bmatrix} \lambda I_n - A & 0_{n \times q} \\ C & 0_q \\ 0_q & I_q \end{bmatrix}$$

$$= rank \begin{bmatrix} \lambda I_n - A \\ C \end{bmatrix} + q,$$

which implies that $[\mathcal{M}\bar{\mathcal{A}}, \bar{C}]$ is detectable if and only if (A, C) is detectable. Obviously, the assumption that A is Schur is unnecessary. Moreover, it is easy to verify that the estimation error equation is $e_{\bar{x}}(k+1) = (\mathcal{M}\bar{\mathcal{A}} - \bar{\mathcal{L}}\bar{C})e_{\bar{x}}(k) + \mathcal{M}\mathcal{F}x_f(k)$ and in this case, $\mathcal{M}\mathcal{F} = \begin{bmatrix} 0 \\ \epsilon_0 I_q \end{bmatrix}$, which implies that the effect of $x_f(k)$ in the estimation error equation can be reduced significantly.

7.4 DELAY COMPENSATION IN CONTROLLER-TO-ACTUATOR NETWORK CHANNEL

In the previous section, the network communication delay in the sensor-to-controller network channel is compensated by regarding the delayed measurement as an output "disturbance," and the EFSO is proposed to accurately estimate the system state and disturbance. Obviously, if the system state can be estimated accurately, the impact of the network communication delay in the sensor-to-controller channel can be eliminated effectively, and the state estimate applied into the controller will be more accurate. In this section, we are in a position to consider the delay compensation problem in the controller-to-actuator channel.

It should be noted that, in the EFSO (7.5), the current input $u(k)$ of the plant plays an important role in obtaining the state estimates, and the operation of the EFSO needs to know the current input $u(k)$. In CPSs, however, as a result of the controller and actuator are distributively located, and there exists time-varying network communication delays in the controller-to-actuator and sensor-to-controller channels, it is complicated and challenging to know the current input of the plant at the controller side. In the following, two possible strategies are proposed to address this problem.

Strategy I: Buffer-based approach. According to Assumption 7.2, the time-varying network communication delay d_k^{ca} in the controller-to-actuator channel is bounded by $\bar{d}_{ca}$. Thus, if the control input signal sent by controller suffer from the delay d_k^{ca}, we can artificially enlarge the delay d_k^{ca} to the bound $\bar{d}_{ca}$ by implementing a buffer at the actuator side, which implies that the time-varying network communication delay d_k^{ca} is converted into a constant one $\bar{d}_{ca}$, thus, enables the controller to know the current control input for the plant since the delay bound $\bar{d}_{ca}$ is known for the controller. Then, at each time step k, we can calculate a $\bar{d}_{ca}$-steps ahead prediction of the control input $u(k + \bar{d}_{ca}|k)$, and send it to the actuator side via network. Due to the controller-to-actuator delay is enlarged as constant delay $\bar{d}_{ca}$, $u(k + \bar{d}_{ca}|k)$ will be applied to the plant at time $k + \bar{d}_{ca}$, which indicates that $u(k + \bar{d}_{ca}) = u(k + \bar{d}_{ca}|k)$, accordingly, we have $u(k) = u(k|k - \bar{d}_{ca})$. Considering that $u(k + \bar{d}_{ca})$ is generated at time k, we can conclude that the control inputs at $u(k + i)$, $i = 0, 1, \cdots, \bar{d}_{ca} - 1$ are produced before time k. The network communication delay in the controller-to-actuator channel, thus, can be compensated. Next, we will consider the construction of the predictive control input $u(k + \bar{d}_{ca}|k)$.

According to the EFSO designed in the previous section, the state estimate $\hat{x}(k)$ can be obtained at time k. Since the input signals from $u(k)$ to $u(k + \bar{d}_{ca} - 1)$ are available in the controller side at time k, we can compute the state estimates $\hat{x}(k+i|k)$, $i = 1, 2, \ldots, \bar{d}_{ca}$, based on these input signals and system (7.1) as follows

$$\hat{x}(k + 1|k) = A\hat{x}(k) + Bu(k)$$
$$\hat{x}(k + 2|k) = A\hat{x}(k + 1|k) + Bu(k + 1)$$
$$\vdots$$
$$\hat{x}(k + \bar{d}_{ca}|k) = A\hat{x}(k + \bar{d}_{ca} - 1|k) + Bu(k + \bar{d}_{ca} - 1).$$

By iterations, we can further obtain that:

$$\hat{x}(k + \bar{d}_{ca}|k) = A^{\bar{d}_{ca}}\hat{x}(k) + \sum_{j=1}^{\bar{d}_{ca}} A^{\bar{d}_{ca}-j}Bu(k + j - 1). \tag{7.14}$$

Hence, the CIP $u(k + \bar{d}_{ca}|k)$ can be designed as

$$u(k + \bar{d}_{ca}|k) = K\hat{x}(k + \bar{d}_{ca}|k) + N_r r(k + \bar{d}_{ca}) \tag{7.15}$$

where the matrix K is the feedback control gain and N_r is the feedforward gain to be designed later.

Strategy II: Packet-based approach. In the previous discussion, the buffer-based approach is proposed to achieve the objective that the observer can exactly know the current control input of the plant. However, it is worth pointing out that, the buffer-based approach is effective when $\bar{d}_{ca}$ is small, which limits the application of the buffer-based control approach. In what follows, the networked delay compensation control approach will be proposed to compensate for the negative effects of the large network communication delay.

According to the buffer-based approach, at time k, the CPIs $u(k+\bar{d}_{ca}-i|k-i)$, $i = 1, 2, \ldots, \bar{d}_{ca}$, have already been determined at time $k-i$, $i = 1, 2, \ldots, \bar{d}_{ca}$, respectively. Then, we can encapsule the previous CPIs $u(k + \bar{d}_{ca} - i|k)$, $i = 1, 2, \ldots, \bar{d}_{ca}$ as well as the current CPI $u(k + \bar{d}_{ca}|k)$ into one "packet," and then send it to the actuator side over network. More specifically, at time k, the CIPs packet $U(k) = [u(k|k - \bar{d}_{ca}), u(k + 1|k - \bar{d}_{ca} + 1), \cdots, u(k + \bar{d}_{ca}|k)]$ will be transmitted to actuator side.

Because of the time-varying network communication delay is bounded by $\bar{d}_{ca}$, one of the CIPs packets $U(k - \bar{d}_{ca})$, $U(k - \bar{d}_{ca} + 1)$, $\cdots$, $U(k)$ is available for the actuator at instant k, but which one will be received at the actuator side at the time k is unknown due to the time-varying network communication delay. Moreover, the packet disorder is also inevitable, thus there may be more than one CIP packets will be received by the actuator, if this is the case, at the actuator side, the actuator always compares the time stamps of both the newly arrived CIPs packet and the one in the buffer, and only the latest CIPs packet is stored, and then the actuator selects the control input signal from the packet according to the current time instant. It should be noted that, according to the construction of the data packet $U(k)$, the control input $u(k) = u(k|k - \bar{d}_{ca})$ chosen for time instant k exists identically in each of the CIPs packets $U(k - \bar{d}_{ca})$, $U(k - \bar{d}_{ca} + 1)$, $\cdots$, $U(k)$. Thus, no matter which packet is available, the control input signal is unique. In this situation, at the time k, the controller can know the current control input signal exactly. According to Strategy I, (7.14) can be verified. Therefore, the CIP $u(k + \bar{d}_{ca}|k)$ can be designed as (7.15).

Remark 7.4. *The main difference between the buffer-based approach and the packet-based approach is that, at time k, only one CIP $u(k + \bar{d}_{ca}|k)$ is to be sent by using buffer-based approach, while the packet-based approach requires not only the current CIP $u(k + \bar{d}_{ca}|k)$ but also the previous CIPs $u(k + \bar{d}_{ca} - i|k - i)$, $i = 1, 2, \ldots, \bar{d}_{ca}$ to be transmitted in a packet.*

7.5 STABILITY AND STEADY-STATE TRACKING ERROR ANALYSIS

In this section, the stability analysis of the closed-loop system is performed.

Theorem 7.3. *Suppose that Assumptions 7.1, 7.2, and 7.3 are satisfied. System (7.1) with $u(k) = u(k|k - \bar{d}_{ca})$ is bounded-input-bounded-output (BIBO) stable if there exist matrix $\bar{L}$, and controller gain K in (7.15) such that $M\bar{A} - \bar{L}\bar{C}$ and $A + BK$ are Schur, respectively.*

Proof. According to system (7.1), and after some iterations, it can be shown that

$$x(k + \bar{d}_{ca}) = A^{\bar{d}_{ca}} x(k) + \sum_{j=1}^{\bar{d}_{ca}} A^{\bar{d}_{ca}-j} Bu(k + j - 1). \tag{7.16}$$

Define the state prediction error as $e_x(k|k-j) = x(k) - \hat{x}(k|k-j)$, $j = 1, 2, \ldots, \bar{d}_{ca}$. Subtracting (7.14) from (7.16) gives that $e_x(k + \bar{d}_{ca}|k) = A^{\bar{d}_{ca}} e_x(k)$. Accordingly, we have $e_x(k|k - \bar{d}_{ca}) = A^{\bar{d}_{ca}} e_x(k - \bar{d}_{ca})$.

Then, the control input signal can be expressed as

$$
\begin{aligned}
u(k) = u(k|k - \bar{d}_{ca}) &= K\hat{x}(k|k - \bar{d}_{ca}) + N_r r(k) \\
&= K(x(k) - e_x(k|k - \bar{d}_{ca})) + N_r r(k) \\
&= Kx(k) - KA^{\bar{d}_{ca}} e_x(k - \bar{d}_{ca}) + N_r r(k).
\end{aligned}
\tag{7.17}
$$

Substituting (7.17) into (7.1), we have

$$
\begin{aligned}
x(k + 1) &= (A + BK)x(k) - BKA^{\bar{d}_{ca}} e_x(k - \bar{d}_{ca}) + BN_r r(k) \\
&= (A + BK)x(k) - D_e e_{\bar{x}}(k - \bar{d}_{ca}) + BN_r r(k),
\end{aligned}
\tag{7.18}
$$

where $D_e = BKA^{\bar{d}_{ca}} [I_n \ \ 0]$.

According to (7.13) and (7.18), the closed-loop system can be written as

$$z(k + 1) = \Omega z(k) + \begin{bmatrix} BN_r \\ 0 \end{bmatrix} r(k),$$

where $z(k) = \begin{bmatrix} x^T(k) & e_{\bar{x}}^T(k - d) \end{bmatrix}^T$ and $\Omega = \begin{bmatrix} A + BK & -D_e \\ 0 & M\bar{A} - \bar{L}\bar{C} \end{bmatrix}$.

Since $r(k)$ is bounded, the closed-loop system is BIBO stable if there exist gain matrices K and $\bar{L}$ such that $A + BK$ and $M\bar{A} - \bar{L}\bar{C}$ are Schur, respectively. This completes the proof. □

Next, the steady-state tracking error of the networked system will be analyzed, and the gain N_r in controller (7.15) will be designed to guarantee the steady tracking error to be zero.

When system (7.1) is at the steady-state, the state, control input, output, and reference input are denoted as x_{ss}, u_{ss}, y_{ss}, and r_{ss}, respectively. Now, assuming the tracking error is zero, i.e., $y_{ss} = Cx_{ss} = r_{ss}$. According to system (7.1), we also have the following relation, $x_{ss} = Ax_{ss} + Bu_{ss}$ or equivalently

$$\Psi \begin{bmatrix} x_{ss} \\ u_{ss} \end{bmatrix} = \begin{bmatrix} 0 \\ r_{ss} \end{bmatrix} \tag{7.19}$$

where $\Psi = \begin{bmatrix} A - I & B \\ C & 0 \end{bmatrix}$. Then, under Assumption 7.1, solving (7.19) yields that,

$$\begin{bmatrix} x_{ss} \\ u_{ss} \end{bmatrix} = \Psi^T (\Psi \Psi^T)^{-1} \begin{bmatrix} 0 \\ r_{ss} \end{bmatrix}$$

$$\triangleq \begin{bmatrix} \Pi_{11} & \Pi_{12} \\ \Pi_{21} & \Pi_{22} \end{bmatrix} \begin{bmatrix} 0 \\ r_{ss} \end{bmatrix} \tag{7.20}$$

which implies that x_{ss} and u_{ss} can be easily solved out from (7.20).

Since $u(k) = K\hat{x}(k|k - \bar{d}_{ca}) + N_r r(k)$, and $M\bar{A} - \bar{L}\bar{C}$ is Schur stable, it follows from (7.13) that the estimation error will be zero at steady-state, and we can eventually replace $x(k)$ by their estimate $\hat{x}(k|k - \bar{d}_{ca})$. Thus, we have $u_{ss} = Kx_{ss} + N_r r_{ss}$ at the steady-state, which together with (7.20) gives that $\Pi_{22}r = K\Pi_{12} + N_r r_{ss}$. Then, N_r can be determined as

$$N_r = \Pi_{22} - K\Pi_{12}. \tag{7.21}$$

Remark 7.5. *Note that N_r is derived by assuming the tracking error is zero. Thus, the necessary condition for the networked systems having zero steady-state tracking error is that the gain matrix K and the coefficient N_r satisfy (7.21).*

Remark 7.6. *It is worth mentioning that the proposed control approaches are established without taking into account the possible uncertainty in system (7.1). However, after some modification, the proposed control approaches are also effective for uncertain system. Consider the following uncertain system,*

$$x(k + 1) = (A + \Delta A(k))x(k) + Bu(k)$$
$$y(k) = Cx(k), \tag{7.22}$$

where A, B, and C are known nominal matrices, $\Delta A(k)$ is an unknown matrix representing the time-varying parameter uncertainties and has the form $\Delta A(k) = \Phi \Sigma(k) \Psi$, Φ and Ψ are known matrices, $\Sigma(k)$ is an unknown time-varying matrix satisfying that $\Sigma^T(k)\Sigma(k) \leq I$. For uncertain system (7.22), we can also design the EFSO (7.5) by using the nominal system to obtain the state estimate $\hat{x}(k)$. Then, similar to the buffer-based approach, after some calculations, we can obtain that

$$\xi(k + 1) = (\mathbf{A} + \Delta \mathbf{A}(k))\,\xi(k) - \mathbf{D}_e e_{\bar{x}}(k - \bar{d}_{ca}) - \mathbf{N}_r r(k),$$

where $\xi(k) = \begin{bmatrix} x^T(k) & x^T(k-1) & \cdots & x^T(k - \bar{d}_{ca}) \end{bmatrix}^T$, and

$$\mathbf{A} = \left[\begin{array}{c|c} A + BK & 0_{n \times n\bar{d}_{ca}} \\ \hline I_{n\bar{d}_{ca}} & 0_{n\bar{d}_{ca} \times n} \end{array} \right], \quad \Delta\mathbf{A}(k) = \begin{bmatrix} BK\bar{\Phi}\bar{\Sigma}(k)\bar{\Psi} \\ 0_{n\bar{d}_{ca} \times (n\bar{d}_{ca}+n)} \end{bmatrix}$$

$$\mathbf{D}_e = \left[\begin{array}{c|c} BKA^{\bar{d}_{ca}} & 0 \\ \hline 0_{n\bar{d}_{ca} \times (n+1)} \end{array} \right], \quad \mathbf{N}_r = \begin{bmatrix} BN_r \\ 0_{n\bar{d}_{ca} \times n} \end{bmatrix}$$

$$\bar{\Phi} = \begin{bmatrix} 0 & \Phi & \cdots & A^{\bar{d}_{ca}-1}\Phi \end{bmatrix}, \quad \bar{\Psi} = diag\{\Psi, \cdots, \Psi\},$$

$$\bar{\Sigma}(k) = diag\left\{\Sigma(k), \cdots, \Sigma(k - \bar{d}_{ca})\right\}.$$

Then, the closed-loop system can be written into the following uncertain system form,

$$\mathbf{z}(k+1) = \mathbf{W}\mathbf{z}(k) + \begin{bmatrix} \mathbf{N}_r \\ 0 \end{bmatrix} r(k)$$

where $\mathbf{z}(k) = \begin{bmatrix} \xi^T(k) & e_{\tilde{x}}^T(k-d) \end{bmatrix}^T$ *and*

$$\mathbf{W} = \begin{bmatrix} \mathbf{A} + \Delta\mathbf{A}(k) & -\mathbf{D}_e \\ 0 & M\bar{A} - \bar{L}\bar{C} \end{bmatrix}.$$

Finally, the stability analysis problem can be solved by using the robust control theory.

7.6 NUMERICAL SIMULATION

In order to verify the effectiveness of the proposed method, we reconsider the servo motor control system in Chapter 5. According to (7.6), we can easily obtain

$$M = \begin{bmatrix} 1 & 0 & 0 \\ 0 & 1 & 0 \\ 0 & 0 & 1 \\ -0.0541 & -0.115 & -0.0001 \end{bmatrix}, \ N = \begin{bmatrix} 0 \\ 0 \\ 0 \\ 1 \end{bmatrix}.$$

Choosing the eigenvalues of matrix $M\bar{A} - \bar{L}\bar{C}$ as 0.9, -0.5213, 0.9963, 0.645, then the pole assignment technique is used to compute the matrix $\bar{L}$ as

$$\bar{L} = \begin{bmatrix} 0.0700 \\ -0.1042 \\ 0.3736 \\ -0.8918 \end{bmatrix}$$

and using (7.8), the observer parameters can be determined as

$$F = \begin{bmatrix} 1.1162 & 0.2049 & -0.3350 & -0.0700 \\ 1.0056 & 0.0120 & 0.0000 & 0.1042 \\ -0.0202 & 0.9570 & -0.0000 & -0.3736 \\ -0.1273 & 0.0909 & 0.0182 & 0.8918 \end{bmatrix}$$

$$G = 10^{-15} \times \begin{bmatrix} 0.0139 \\ -0.0139 \\ 0.0555 \\ -0.1110 \end{bmatrix}, H = \begin{bmatrix} 1.0000 \\ 0.0000 \\ 0.0000 \\ -0.0541 \end{bmatrix}, Q = N.$$

Similarly, by selecting the eigenvalues of matrix $A + BK$ as 0.41, 0.72, 0.53, the feedback control gain matrix K can be chosen as $K = \begin{bmatrix} 0.5400 & -1.1071 & 0.4915 \end{bmatrix}$. Accordingly, using (7.21) and Ψ with Π_{12}, Π_{22}, the coefficient N_r can be computed as $N_r = 0.4589$.

To illustrate the effectiveness of the proposed method in this chapter, we assume that the plant and the controller are connected via a communication network, i.e., the output and input signals are transmitted via network. In this case, the time-varying communication delays in the controller-to-actuator and sensor-to-controller channels are assumed to be less than 0.28s, respectively. As the sample time is 0.04s, the upper bounds are $\bar{d}_{ca} = 7$ and $\bar{d}_{sc} = 7$. The initial conditions are $x(0) = [0, 0, 0]^T$, $\hat{\bar{x}}(0) = [0, 0, 0, 0]^T$, and $u(k) = 0$, $k = 0, 1, 2, \ldots, \bar{d}_{ca}$. The desired signals to be tracked by $y(k)$ are

$$r_1(k) = 3, \ k \geq 0$$

$$r_2(k) = \begin{cases} 3 & \text{if } 0 \leq k \leq 100 \text{ and } 200 \leq k \leq 300 \\ -3 & \text{if } 100 \leq k \leq 200 \text{ and } k > 300 \end{cases}$$

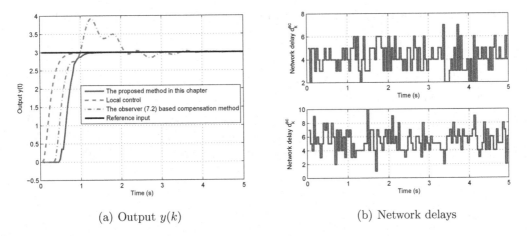

(a) Output $y(k)$ (b) Network delays

Figure 7.2 Comparative outputs of the motor control system and network delays $(r(k) = r_1(k))$

For the sake of comparison, the traditional observer (7.2) is also applied to the system and the simulation results are shown in Figures 7.2(a) and 7.3(a), where the traditional observer is used to generate the one step ahead state estimate $\hat{x}(k + 1|k)$ without considering the delay compensation in the sensor-to-controller channel, and the packet-based approach is applied to compensate for the network delay in the controller-to-actuator channel. From the simulation results, it can be seen that the response and the overshoot of the system with the proposed packet-based approach in this chapter are obviously faster than that applying the traditional observer. In the simulation, the network delays in sensor-to-controller and controller-to-actuator channels are generated randomly with the upper bound $\bar{d}_{ca} = 7$, $\bar{d}_{sc} = 7$, and shown in Figure 7.2(b) and Figure 7.3(b), respectively.

7.7 SUMMARY

In this chapter, the output tracking control problem has been investigated by considering the delay compensation in both controller-to-actuator and sensor-to-controller

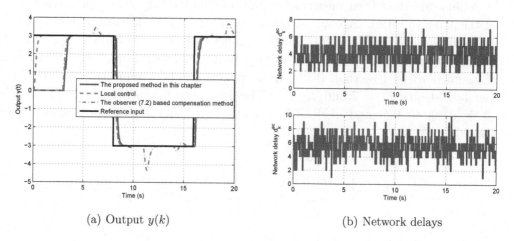

(a) Output $y(k)$ (b) Network delays

Figure 7.3 Comparative outputs of the motor control system and network delays $(r(k) = r_2(k))$

channels. The EFSO has been designed to compensate the sensor-to-controller channel delay by viewing the delayed measurement as a special disturbance, and the buffer- and packet-based delay compensation control approach have been presented to compensate the controller-to-actuator channel delay. Finally, a DC-motor system has been used to show the effectiveness of the proposed control methods.

II

Event-Triggered Control Approaches in CPSs

II

Event-Triggered Control Approaches in CPSs

CHAPTER 8

Observer-Based Event-Triggered Control for CPSs

8.1 INTRODUCTION

In CPSs, [171, 211], sensors communicate with controllers through a load or bandwidth limited network, and in such a case, communication between sensors and controllers should be as little as possible to avoid congestion or packet dropouts. To cope with this problem, event-triggered control approaches have been proposed as the means to reduce the communication frequency between the components of the control systems. In event-triggered control framework, the necessary updating or communication is determined by the occurrence of an "event" rather than "time." It should be pointed out that the event-triggered control approaches in [229, 223] are developed under the assumption that all the internal plant states are available for controller design. However, for many practical applications, the full state information is not always available, and only the output is accessible. One solution to address this problem is to use a state observer to reconstruct the system states. In [194], the observer-based event-triggered consensus problem is considered for multi-agent systems with linear dynamics, and both centralized and distributed event-triggered cooperative control strategies are proposed. In [85], the functional observer based event-triggered controller design is considered for linear systems, two approaches are proposed for the event-triggered control system by applying a hybrid system and interpreting the event-induced error as exogenous disturbance. In [119], the event-triggered non-PDC control problem is investigated for networked T-S fuzzy systems with limited data transmission bandwidth and the imperfect premise matching membership functions. In [80], the event-triggered extended state observer is designed for a continuous-time nonlinear system with uncertainty and disturbance, an event-triggered transmission strategy is proposed, and the observation error is uniformly bounded.

This chapter concerns the problem of observer-based event-triggered output feedback control of linear systems. Contrary to normal sampled-data control systems, the controller, in this chapter, is updated only when an "event" happens, and a typical

DOI: 10.1201/9781003260882-8

event is defined as some error signals exceeding a given threshold. Both continuous and discrete updating instants scheduler (UIS) cases are considered. It is shown that even with the significantly reduced updating frequency of the controller, the global uniform ultimate boundedness of the states of the event-triggered closed-loop systems can also be guaranteed. Numerical example is finally used to illustrate the effectiveness and advantages of the proposed approaches.

8.2 SYSTEM DESCRIPTION

As shown in Figure 8.1, the event-triggered control system considered in this chapter can be grouped into the following three modules: (1) the physical plant and smart sensor, (2) the updating instants scheduler (UIS), and (3) the event-triggered controller.

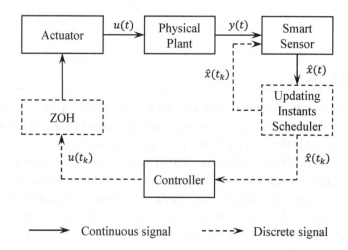

Figure 8.1 Block diagram of event-triggered control system

The physical plant is given by the following continuous-time linear system:

$$\dot{x}(t) = Ax(t) + Bu(t)$$
$$y(t) = Cx(t) \tag{8.1}$$

where $x(t) \in \mathbb{R}^n$ represents the system state vector, $u(t) \in \mathbb{R}^m$ denotes the control input vector, and $y(t) \in \mathbb{R}^q$ is the output vector, and A, B, C are system matrices with appropriate dimensions. It is assumed that the pairs (A, B) and (C, A) are controllable and observable, respectively.

Here, we assume that the sensor has necessary computation capability, and it can pre-process measurement $y(t)$ to obtain the state estimate $\hat{x}(t)$ according to the following state observer:

$$\dot{\hat{x}}(t) = A\hat{x}(t) + Bu(t) + L(y(t) - C\hat{x}(t)), \tag{8.2}$$

where $\hat{x}(t) \in \mathbb{R}^n$ is the observer state, L is the observer gain appropriately designed. With the state estimate $\hat{x}(t)$, the event-triggered controller is designed as

$$u(t) = K\hat{x}(t_k), \ t \in [t_k, t_{k+1}), \ k \in \mathbb{N} \tag{8.3}$$

where K is the controller gain with appropriate dimension, t_k is the updating instant determined by the UIS, which monitors the event-triggering condition to determine whether an "event" is generated or not. Once an event happens, the UIS will transmit the newest state estimate to the event-triggered controller. More specifically, let us denote the time instants when an event happens by $\{t_k\}_{k=0}^{\infty}$ with $t_k < t_{k+1}$. Without loss generality, we assume that $t_0 = 0$, and the first event is generated at time t_0.

Notice that in the event-triggered control scheme, the controller in (8.3) will receive the sampled estimated state $\hat{x}(t_k)$ at sampling instant t_k, and $\hat{x}(t_k)$ will be held until next event happens at time t_{k+1}. In this situation, the control input (8.3) is only computed at the sampling instants. To keep the control signal continuous, a zero-order hold (ZOH) is embedded.

In this chapter, the problem of observer-based event-triggered output feedback control will be addressed. More specifically, we will design the UISs to determine the updating instant t_k, and also show that there exists a minimum inter-event interval defined by $t_{\min} = \inf_{k \in \mathbb{N}} \{t_{k+1} - t_k\}$ to exclude the Zeno behavior. Then, with the defined updating instants, the stability of the closed-loop system is analyzed.

8.3 EVENT-TRIGGERED CONTROL WITH CONTINUOUS UIS

With the estimated states $\hat{x}(t)$, the next updating instant t_{k+1} can be determined by

$$t_{k+1} = \inf \{t > t_k | \|e(t)\| \geq \gamma(t)\}, \ \forall k \in \mathbb{N} \tag{8.4}$$

where $e(t) = \hat{x}(t) - \hat{x}(t_k)$, $\gamma(t) = \sqrt{\varepsilon e^{-\alpha t} + \varepsilon_0}$ is the exponentially decreasing event threshold with given parameters $\varepsilon > 1$, $0 \leq \alpha < 1$, and $\varepsilon_0 \geq 0$. It is clear that, in the UIS (8.4), the state estimate $\hat{x}(t)$ is detected continuously.

Define the estimation error $\tilde{x}(t) = x(t) - \hat{x}(t)$. For the time period $t \in [t_k, t_{k+1})$, the plant and error system can be rewritten as

$$\begin{aligned}
\dot{x}(t) &= Ax(t) + BK\hat{x}(t_k) \\
&= (A + BK)x(t) - BK\tilde{x}(t) - BK\left(\hat{x}(t) - \hat{x}(t_k)\right) \\
&= (A + BK)x(t) - BK\tilde{x}(t) - BKe(t),
\end{aligned} \tag{8.5}$$

where $e(t) = \hat{x}(t) - \hat{x}(t_k)$, and

$$\dot{\tilde{x}}(t) = (A - LC)\tilde{x}(t). \tag{8.6}$$

Define $\xi(t) = [x^T(t), \tilde{x}^T(t)]^T$, we have the following closed-loop system

$$\dot{\xi}(t) = \bar{A}\xi(t) + \bar{B}e(t), \ t \in [t_k, t_{k+1}) \tag{8.7}$$

where $\bar{A} = \begin{bmatrix} A + BK & -BK \\ 0 & A - LC \end{bmatrix}$, $\bar{B} = \begin{bmatrix} -BK \\ 0 \end{bmatrix}$.

The following theorem shows that the global uniform ultimate boundedness of the states of the closed-loop system (8.7) can be established.

Theorem 8.1. *Consider the closed-loop system (8.7) with updating instants determined by (8.4). For given gain matrices K and L, if there exists a symmetric and positive definite matrix $P = P^T > 0$ satisfying the following matrix inequality,*

$$Q \triangleq \bar{A}^T P + P \bar{A} + P \bar{B} \bar{B}^T P < 0, \tag{8.8}$$

then the states of system (8.7) are globally UUB, and exponentially converge to the bounded region $\mathcal{B}(\varepsilon_0) = \left\{ \xi(t) | \|\xi(t)\| \leq \sqrt{\frac{\varepsilon_0}{\delta \lambda_{\min}(P)}} \right\}$, where $\delta = \frac{\lambda_{\min}(-Q)}{\lambda_{\max}(P)}$.

Proof. Considering the Lyapunov function candidate $V(t) = \xi^T(t) P \xi(t)$ with symmetric and positive definite matrix $P = P^T > 0$ being a solution of (8.8), the time derivative of $V(t)$ for $t \in [t_k, t_{k+1})$ is

$$\begin{aligned}
\dot{V}(t) &= \dot{\xi}^T(t) P \xi(t) + \xi^T(t) P \dot{\xi}(t) \\
&= \left(\bar{A} \xi(t) + \bar{B} e(t) \right)^T P \xi(t) + \xi^T(t) P \left(\bar{A} \xi(t) + \bar{B} e(t) \right) \\
&= \xi^T(t) \left(\bar{A}^T P + P \bar{A} \right) \xi(t) + e^T(t) \bar{B}^T P \xi(t) + \xi^T(t) P \bar{B} e(t) \\
&= \xi^T(t) \left(\bar{A}^T P + P \bar{A} + P \bar{B} \bar{B}^T P \right) \xi(t) - \left\| e(t) - \bar{B}^T P \xi(t) \right\|^2 + \|e(t)\|^2 \\
&\leq -\lambda_{\min}(-Q) \|\xi(t)\|^2 + \|e(t)\|^2.
\end{aligned}$$

According to the updating instants defined in (8.4), we can find that, as long as $t \in [t_k, t_{k+1})$, $\|e(t)\| \leq \gamma(t)$ holds. Hence, $\forall t \in [t_k, t_{k+1})$, we have

$$\begin{aligned}
\dot{V}(t) &\leq -\lambda_{\min}(-Q) \|\xi(t)\|^2 + \gamma^2(t) \\
&\leq -\frac{\lambda_{\min}(-Q)}{\lambda_{\max}(P)} \xi^T(t) P \xi(t) + \gamma^2(t) \\
&= -\delta V(t) + \gamma^2(t). \tag{8.9}
\end{aligned}$$

Applying the comparison lemma [86] to (8.9) yields

$$V(t) \leq e^{-\delta t} V(0) + \int_0^t e^{-\delta(t-s)} \gamma^2(s) \mathrm{d}s. \tag{8.10}$$

Note that both $V(t)$ and $\gamma(t)$ are continuous for all $t \geq 0$, the inequality (8.10) always holds for all $t \geq 0$.

Noticing that $\gamma^2(t) = \varepsilon^{-\alpha t} + \varepsilon_0 = e^{-(\alpha \ln \varepsilon)t} + \varepsilon_0$, we can obtain the following inequality from (8.10),

$$\begin{aligned}
V(t) &\leq e^{-\delta t} V(0) + e^{-\delta t} \int_0^t e^{(\delta - \alpha \ln \varepsilon)s} \mathrm{d}s + \int_0^t e^{-\delta(t-s)} \varepsilon_0 \mathrm{d}s \\
&= e^{-\delta t} \left(V(0) - \frac{\varepsilon_0}{\delta} \right) + \frac{\varepsilon_0}{\delta} + e^{-\delta t} \int_0^t e^{(\delta - \alpha \ln \varepsilon)s} \mathrm{d}s.
\end{aligned}$$

In what follows, three cases will be considered. If $\delta - \alpha \ln \varepsilon = 0$, we obtain

$$V(t) \le e^{-\delta t} \left(V(0) - \frac{\varepsilon_0}{\delta} + t \right) + \frac{\varepsilon_0}{\delta}.$$

If $\delta - \alpha \ln \varepsilon > 0$, we have

$$V(t) \le e^{-\delta t} \left(V(0) - \frac{\varepsilon_0}{\delta} \right) + \frac{\varepsilon_0}{\delta} + \frac{e^{-\delta t}}{\delta - \alpha \ln \varepsilon} \left(e^{(\delta - \alpha \ln \varepsilon)t} - 1 \right)$$

$$= e^{-\delta t} \left(V(0) - \frac{\varepsilon_0}{\delta} - \frac{1}{\delta - \alpha \ln \varepsilon} \right) + \frac{\varepsilon_0}{\delta} + \frac{\varepsilon^{-\alpha t}}{\delta - \alpha \ln \varepsilon}.$$

If $\delta - \alpha \ln \varepsilon < 0$, then $e^{(\delta - \alpha \ln \varepsilon)t} < 1$, similarly, we have

$$V(t) \le e^{-\delta t} \left(V(0) - \frac{\varepsilon_0}{\delta} - \frac{1}{\delta - \alpha \ln \varepsilon} \right) + \frac{\varepsilon_0}{\delta} + \frac{\varepsilon^{-\alpha t}}{\alpha \ln \varepsilon - \delta}.$$

Thus, we can conclude that, no matter what the value of $\delta - \alpha \ln \varepsilon$ is, the global uniform ultimate boundedness of the states of the resulting closed-loop system (8.7) can be guaranteed, and the states exponentially converge to a bounded region $\mathcal{B}(\varepsilon_0)$. The proof is thus completed. □

Remark 8.1. *In Theorem 8.1, with the UIS (8.4), the global uniform ultimate boundedness of the states of system (8.7) is guaranteed for the observer-based event-triggered control system by using an exponentially decreasing event triggering condition. It is clear that, if we choose $\alpha = 0$, the event triggering condition (8.4) reduces into the constant event triggering condition considered in [106]. Furthermore, if we choose $\varepsilon_0 = 0$, the global asymptotic stability can also be obtained.*

Remark 8.2. *It can be seen that the bound $\mathcal{B}(\varepsilon_0)$ is independent of the parameters ε and α, which indicates that it is possible to reduce the updating times during the transient-state by appropriately selecting the parameters ε and α. At the same time, we can choose ε_0 small enough to guarantee better control performance. This is not possible for the approaches based on the constant event threshold considered in [106, 90], where more updating times are required during the transient-state.*

To exclude the Zeno behavior of the updating, we need to show there always exists a nonzero lower bound of the minimum inter-event interval. Let's first recall the following lemma.

Remark 8.3. *[211] Assume $A \in \mathbb{R}^{n \times n}$ is Hurwitz, then there exists a positive scalar $c > 0$ such that $\|e^{At}\| \le c e^{\frac{\lambda_{\max}(A)}{2} t}$, where $\lambda_{\max}(A) = \max_i \{\text{Re}(\lambda_i(A))\}$.*

The following theorem indicates that there exists a lower bound of the minimum inter-event interval.

Theorem 8.2. *With the updating instants determined by (8.4), the minimum inter-event interval is lower bounded by a positive scalar.*

Proof. For any $k \in \mathbb{N}$, let t_k be an updating instant. Then, in the time interval $[t_k, t_{k+1})$, $\hat{x}(t_k)$ is constant. According to the definition of $e(t)$ in (8.5), for $t \in [t_k, t_{k+1})$, we have $\dot{e}(t) = Ae(t) + (A + BK)\hat{x}(t_k) + LC\tilde{x}(t)$, which yields that

$$e(t) = e^{A(t-t_k)}e(t_k) + \int_{t_k}^{t} e^{A(t-s)}\left((A + BK)\hat{x}(t_k) + LC\tilde{x}(s)\right)ds$$

$$= \int_{t_k}^{t} e^{A(t-s)}\left((A + BK)\hat{x}(t_k) + LC\tilde{x}(s)\right)ds.$$

Moreover, noticing (8.6), we have $\dot{\tilde{x}}(t) = (A - LC)\tilde{x}(t)$ which leads to $\tilde{x}(t) = e^{(A-LC)t}\tilde{x}(0)$, where $\tilde{x}(0)$ is the initial estimate error, and it is natural to assume that $\|\tilde{x}(0)\|$ is bounded.

Then, we have

$$\|e(t)\| = \left\|\int_{t_k}^{t} e^{A(t-s)}[(A + BK)\hat{x}(t_k) + LC\tilde{x}(s)]ds\right\|$$

$$\leq \int_{t_k}^{t} e^{\|A\|(t-s)}\|(A + BK)\hat{x}(t_k) + LC\tilde{x}(s)\|ds$$

$$\leq \int_{t_k}^{t} e^{\|A\|(t-s)}(\|(A + BK)\hat{x}(t_k)\| + \|LC\|\|\tilde{x}(s)\|)ds$$

$$\leq \int_{t_k}^{t} e^{\|A\|(t-s)}(\|A + BK\|\|\hat{x}(t_k)\| + \|LC\|\left\|e^{(A-LC)s}\right\|\|\tilde{x}(0)\|)ds.$$

With the aid of Lemma 8.3 and noticing $A - LC$ is Hurwitz with $\lambda_{\max}(A - LC) < 0$, we have

$$\|e(t)\| \leq \int_{t_k}^{t} e^{\|A\|(t-s)}(\|A + BK\|\|\hat{x}(t_k)\| + ce^{\frac{\lambda_{\max}(A-LC)}{2}s}\|LC\|\|\tilde{x}(0)\|)ds$$

$$\leq \phi(t_k)\int_{t_k}^{t} e^{\|A\|(t-s)}ds,$$

where $\phi(t_k) = \|A + BK\|\|\hat{x}(t_k)\| + ce^{\frac{\lambda_{\max}(A-LC)}{2}t_k}\|LC\|\|\tilde{x}(0)\|$.

If $\|A\| \neq 0$, we have $\|e(t)\| \leq \frac{\phi(t_k)}{\|A\|}\left(e^{\|A\|(t-t_k)} - 1\right)$. According to the definition of updating instants (8.4), the next event will not be generated before $\|e(t)\| = \gamma(t)$. Therefore, a lower bound on the inter-event interval denoted by $\bar{T} = t - t_k$ can be determined by

$$\frac{\phi(t_k)}{\|A\|}\left(e^{\|A\|\bar{T}} - 1\right) = \sqrt{\varepsilon^{-\alpha(\bar{T}+t_k)} + \varepsilon_0} \tag{8.11}$$

which means that for any given updating instant t_k, $\bar{T}$ cannot be zero, thus $\bar{T} > 0$.

Similarly, if $\|A\| = 0$, we have $\|e(t)\| \leq \phi(t_k)(t - t_k)$ and the lower bound $\bar{T}$ can be computed by $\bar{T}\phi(t_k) = \sqrt{\varepsilon^{-\alpha(\bar{T}+t_k)} + \varepsilon_0}$, which also indicates that $\bar{T} > 0$.

With the above discussion, it can be concluded that there exists a positive lower bound of the minimum inter-event interval $t_{\min}$. This completes the proof. □

Remark 8.4. *In Theorem 8.2, the existence of the lower bound of the minimum inter-event interval $t_{\min}$ has been shown. Actually, for each time instant t_k, the lower bound of $t_{\min}$ can be determined analytically. If $\|A\| \neq 0$, from (8.11), it can be shown that $\frac{\phi(t_k)}{\|A\|}\left(e^{\|A\|\bar{T}} - 1\right) \geq \sqrt{\varepsilon_0}$. One thus has, $\bar{T} \geq \frac{1}{\|A\|}\ln\left(1 + \frac{\sqrt{\varepsilon_0}\|A\|}{\phi(t_k)}\right)$. If $\|A\| = 0$, it is easy to obtain that $\bar{T} \geq \frac{\sqrt{\varepsilon_0}}{\phi(t_k)}$.*

8.4 EVENT-TRIGGERED CONTROL WITH DISCRETE-TIME EVENT DETECTOR

It should be noted that the continuous UISs rely on continuous detection of the event triggering condition. To reduce the detection times, we also consider the discrete UISs case. In this case, the UIS works in a discrete manner, i.e., it monitors the event-triggering condition periodically. Here, we assume that the UIS samples the state estimate $\hat{x}(t)$ with constant period T_s. In this situation, the updating instants can be determined by $t_k = i_k T_s$, where i_k, $k \in \mathbb{N}$, are some integers and $\{i_0, i_1, i_2, \ldots\} \subset \{0, 1, 2, 3, \ldots\}$ with $i_0 = 0$ and $i_k < i_{k+1}$. It is obvious that, for $\forall k$, $t_{k+1} - t_k \geq T_s$ holds. Define $l_{k,j} = (i_k + j)T_s$, $j = 0, 1, 2, \ldots, d_k$, and $d_k = i_{k+1} - i_k - 1$. It is obvious that $[t_k, t_{k+1}) = \cup_{j=0}^{d_k}[l_{k,j}, l_{k,j+1})$. Now, we assume $\hat{x}(t_k)$ is sampled for feedback control at updating instant t_k. Define $e_d(t) = \hat{x}(l_{k,j}) - \hat{x}(t_k)$, $t \in [l_{k,j}, l_{k,j+1})$, thus $e_d(t)$ is a piecewise constant and continuous from the right function, the next updating instant t_{k+1} can be given by $t_{k+1} = i_{k+1}T_s$, where

$$i_{k+1} = \min_{h \in \mathbb{Z}} \{h > i_k \,|\, \|e_d(hT_s)\| > \gamma((h+1)T_s)\}, \tag{8.12}$$

where $\gamma(t)$ is given in (8.4).

Now, we consider the discrete UIS case. In this case, the observer (8.2) is modified as

$$\dot{\hat{x}}(t) = A\hat{x}(t) + Bu(t) + L(y(l_{k,j}) - C\hat{x}(l_{k,j})), \tag{8.13}$$

thus, the observer only uses sampled measurements to obtain the state estimate.

Define $\tau(t) = t - l_{k,j}$, thus, for $t \in [l_{k,j}, l_{k,j+1})$, we have

$$\begin{aligned}
\dot{x}(t) &= Ax(t) + BK\hat{x}(t_k) \\
&= Ax(t) + BKx(l_{k,j}) - BKx(l_{k,j}) + BK\hat{x}(l_{k,j}) \\
&\quad - BK\hat{x}(l_{k,j}) + BK\hat{x}(t_k) \\
&= Ax(t) + BKx(t - \tau(t)) - BK\tilde{x}(t - \tau(t)) - BK\left(\hat{x}(l_{k,j}) - \hat{x}(t_k)\right) \\
&= Ax(t) + BKx(t - \tau(t)) - BK\tilde{x}(t - \tau(t)) - BKe_d(t),
\end{aligned}$$

and

$$\dot{\tilde{x}}(t) = A\tilde{x}(t) - LC\tilde{x}(t - \tau(t)). \tag{8.14}$$

Note that $e_d(t)$ is piecewise constant and continuous from the right over time interval $[t_k, t_{k+1})$, and inequality $0 \leq \tau(t) \leq T_s$ holds for $t \in [l_{k,j}, l_{k,j+1})$, $k \in \mathbb{Z}$.

With (8.6), we have the following closed-loop system

$$\dot{\xi}(t) = \bar{A}_d \xi(t) + \bar{A}_\tau \xi(t - \tau(t)) + \bar{B} e_d(t), t \in [l_{k,j}, l_{k,j+1}) \tag{8.15}$$

where $\xi(t)$ was defined as in (8.7), and $\bar{A}_d = \begin{bmatrix} A & 0 \\ 0 & A \end{bmatrix}$, $\bar{A}_\tau = \begin{bmatrix} BK & -BK \\ 0 & -LC \end{bmatrix}$.

In this section, we will perform the stability analysis of the closed-loop system (8.15). To this end, we need to determine the updating instants first. The updating instants considered for discrete UIS case are different from the ones for continuous UIS case since the event-triggering condition is checked periodically. It is clearly that, the inequality $\|e_d(t)\| \leq \gamma(t)$ is always satisfied for $t \in [t_k, t_{k+1})$.

Theorem 8.3. *Consider the closed-loop system (8.15) with updating instants determined by*

$$t_{k+1} = i_{k+1} T_s \tag{8.16}$$

where T_s is the sampling period of the UIS, and i_k is given by (8.12) with $\varepsilon > 1$ and $0 < \alpha < 1$. For given decay rate σ and gain matrices K and L, if there exist matrices $P > 0$, $Z > 0$, $X \geq 0$, M_1, M_2, N_1, N_2 with appropriate dimensions, satisfying the following matrix inequalities,

$$\Omega_l = \Omega_{1l} + \Omega_2 + \Omega_2^T + \Omega_3 < 0, \tag{8.17}$$

$$\Omega_u = \Omega_{1u} + \Omega_2 + \Omega_2^T + \Omega_3 + T_s X < 0, \tag{8.18}$$

$$\Theta = \begin{bmatrix} X & M^T \\ * & e^{-\sigma T_s} Z \end{bmatrix} \geq 0, \tag{8.19}$$

where

$$\Omega_{1l} = \begin{bmatrix} T_s Z & P & 0 \\ P & \sigma P & 0 \\ 0 & 0 & 0 \end{bmatrix}, \quad \Omega_{1u} = \begin{bmatrix} 0 & P & 0 \\ P & \sigma P & 0 \\ 0 & 0 & 0 \end{bmatrix},$$

$$\Omega_2 = \begin{bmatrix} -N^T & M^T + N^T \bar{A}_d & N^T \bar{A}_\tau - M^T \end{bmatrix},$$

$$\Omega_3 = N^T \bar{B} \bar{B}^T N, \quad M = [0, \ M_1, \ M_2], \quad N = [N_1, \ N_2, \ 0],$$

then, the solutions of system (8.15) is globally UUB and exponentially converge to the bounded region $\mathcal{B}_d(\varepsilon_0) = \left\{ \xi(t) \| \|\xi(t)\| \leq \sqrt{\frac{\varepsilon_0}{\sigma \lambda_{\min}(P)}} \right\}$.

Proof. Consider the following time-dependent Lyapunov functional for $t \in [l_{k,j}, l_{k,j+1})$,

$$V(t) = \xi^T(t) P \xi(t) + (T_s - \tau(t)) \int_{t-\tau(t)}^t e^{\sigma(s-t)} \dot{\xi}^T(s) Z \dot{\xi}(s) \mathrm{d}s.$$

Note that $\dot{\tau}(t) = 1$ and $\dot{\xi}(t - \tau(t)) = 0$, we have

$$\dot{V}(t) \leq \xi^T(t)P\dot{\xi}(t) + \dot{\xi}^T(t)P\xi(t) - e^{-\sigma T_s}\int_{t-\tau(t)}^{t}\dot{\xi}^T(s)Z\dot{\xi}(s)ds$$
$$- \sigma(T_s - \tau(t))\int_{t-\tau(t)}^{t}e^{\sigma(s-t)}\dot{\xi}^T(s)Z\dot{\xi}(s)ds$$
$$+ (T_s - \tau(t))\dot{\xi}^T(t)Z\dot{\xi}(t). \tag{8.20}$$

Define $\eta(t) = \left[\dot{\xi}^T(t),\ \xi^T(t),\ \xi^T(t - \tau(t))\right]^T$, it is easy to see that the following facts are true:

$$2\eta^T(t)M^T\left(\xi(t) - \xi(t - \tau(t)) - \int_{t-\tau(t)}^{t}\dot{\xi}(s)ds\right) = 0, \tag{8.21}$$

$$2\eta^T(t)N^T\left(\bar{A}_d\xi(t) + \bar{A}_\tau\xi(t - \tau(t)) - \dot{\xi}(t) + \bar{B}e_d(t)\right) = 0, \tag{8.22}$$

$$\tau(t)\eta^T(t)X\eta(t) - \int_{t-\tau(t)}^{t}\eta^T(t)X\eta(t)ds = 0. \tag{8.23}$$

Adding the terms (8.21)–(8.23) on the righthand side of (8.20), and noticing the following inequality,

$$2\eta^T(t)N^T\bar{B}e_d(t) = \eta^T(t)N^T\bar{B}\bar{B}^TN\eta(t) + \|e_d(t)\|^2 - \left\|e_d(t) - \bar{B}^TN\eta(t)\right\|^2$$
$$\leq \eta^T(t)N^T\bar{B}\bar{B}^TN\eta(t) + \|e_d(t)\|^2.$$

we obtain

$$\dot{V}(t) + \sigma V(t) \leq \eta^T(t)\Omega\eta(t) - \int_{t-\tau(t)}^{t}\zeta^T(t,s)\Theta\zeta(t,s)ds + \|e_d(t)\|^2,$$

where $\zeta(t,s) = \left[\eta^T(t),\ \dot{\xi}^T(s)\right]^T$ and $\Omega = \Omega_1 + \Omega_2 + \Omega_2^T + \Omega_3 + \tau(t)X$ with

$$\Omega_1 = \begin{bmatrix} (T_s - \tau(t))Z & P & 0 \\ P & \sigma P & 0 \\ 0 & 0 & 0 \end{bmatrix}.$$

Note that $\Omega = \frac{T_s-\tau(t)}{T_s}\Omega_l + \frac{\tau(t)}{T_s}\Omega_u$, thus, (8.17) and (8.18) implies that $\Omega < 0$, which together with (8.19) gives that $\dot{V}(t) \leq -\sigma V(t) + \|e_d(t)\|^2$.

Noting that the definition of the updating instants and the inequality $\|e_d(t)\| \leq \gamma(t)$ for $t \in [t_k, t_{k+1})$, one has $\dot{V}(t) \leq -\sigma V(t) + \varepsilon^{-\alpha t} + \varepsilon_0$. Then, following similar arguments of the proof of Theorem 8.1, the global uniform ultimate boundedness of system (8.15) can be established. This completes the proof. □

Remark 8.5. *The conditions (8.17)–(8.19) include the sampling period T_s of the UIS, thus, the admissible sampling period T_s should be selected as the one such that the conditions (8.17)–(8.19) are feasible.*

Remark 8.6. *It can be seen that the conditions (8.17)–(8.19) are dependent on the decay rate σ. In general, the decay rate σ should be chosen as small as possible so that the feasibility of those LMIs will be enhanced. However, with small σ, the transient performance of the closed-loop control systems will be adversely affected.*

Remark 8.7. *It is worth mentioning that, for both continuous and discrete UIS cases, the full-order observer is implemented to estimate the state variables. However, in many cases, part of state variables can be either measured directly or calculated easily from the output. Then, we can adopt a reduced-order observer to estimate the unmeasurable states, which can reduce the real-time computational burdens.*

8.5 NUMERICAL SIMULATION

Consider an inverted pendulum system with a car as shown in Figure 8.2, where M is the mass of the cart, m is the mass of the pendulum, l is the length of the pendulum, x is the cart position coordinate, θ is the pendulum angle from vertical, and F is the input force.

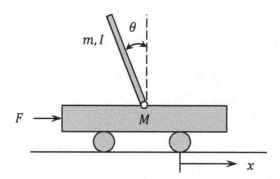

Figure 8.2 Inverted pendulum system

The linearized system model can be represented in state-space form as:

$$
\begin{bmatrix} \dot{x} \\ \ddot{x} \\ \dot{\theta} \\ \ddot{\theta} \end{bmatrix} = \begin{bmatrix} 0 & 1 & 0 & 0 \\ 0 & \frac{-(I+ml^2)b}{I(M+m)+Mml^2} & \frac{m^2gl^2}{I(M+m)+Mml^2} & 0 \\ 0 & 0 & 0 & 1 \\ 0 & \frac{-mlb}{I(M+m)+Mml^2} & \frac{mgl(M+m)}{I(M+m)+Mml^2} & 0 \end{bmatrix} \begin{bmatrix} x \\ \dot{x} \\ \theta \\ \dot{\theta} \end{bmatrix} + \begin{bmatrix} 0 \\ \frac{I+ml^2}{I(M+m)+Mml^2} \\ 0 \\ \frac{ml}{I(M+m)+Mml^2} \end{bmatrix} u,
$$

$$
y = \begin{bmatrix} 1 & 0 & 0 & 0 \\ 0 & 0 & 1 & 0 \end{bmatrix} \begin{bmatrix} x \\ \dot{x} \\ \theta \\ \dot{\theta} \end{bmatrix},
$$

where b is the friction of the cart, I is the inertia of the pendulum. For this example, let's assume that $M = 0.5$ kg, $m = 0.5$ kg, $b = 0.1$ N/m/sec, $l = 0.3$ m, $I = 0.006$ kg·m^2. The initial condition of the system is $[0.98,\ 0,\ 0.2,\ 0]^T$.

We first examine the effectiveness of the event-triggered control approach for the continuous UIS case. By using pole assignment technique, one has the control and observer gains respectively as follows,

$$K = \begin{bmatrix} 17.0386 & 13.0877 & -50.0520 & -9.8150 \end{bmatrix}, L = \begin{bmatrix} 11.0993 & -0.0991 \\ 29.7908 & 2.1306 \\ -0.5518 & 11.3189 \\ -5.5941 & 63.2481 \end{bmatrix}.$$

It can be verified that the matrix inequality (8.8) is feasible for a positive definite matrix P.

Taking $\varepsilon = \mathrm{e}$, $\alpha = 0.034$, $\varepsilon_0 = 0.01$, and under the event triggering condition (8.4), the simulation results are shown in Figure 8.3, where the initial condition of the observer is chosen as $[0.1,\ 0,\ 0,\ 0]^T$.

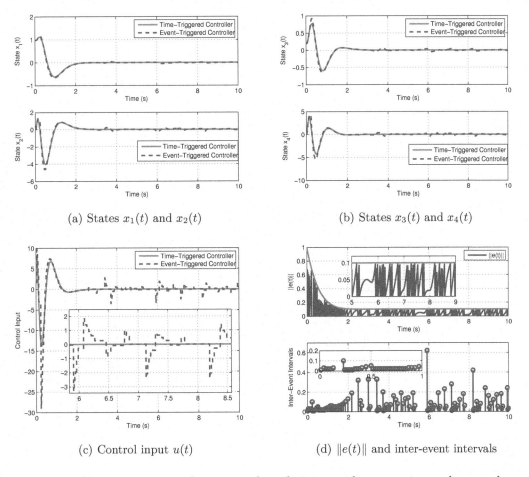

(a) States $x_1(t)$ and $x_2(t)$

(b) States $x_3(t)$ and $x_4(t)$

(c) Control input $u(t)$

(d) $\|e(t)\|$ and inter-event intervals

Figure 8.3 Comparative simulation results of time- and event-triggered control approaches (continuous UIS case)

Figures 8.3(a) and 8.3(b) present the state trajectories of the closed-loop system with event-triggered controller (8.3) and continuous controller $u(t) = K\hat{x}(t)$, respectively. It can be seen that the states of the closed-loop system is globally UUB, and the performance of closed-loop system with the event-triggered controller is almost the same as that of the continuous controller. The control input signal and $\|\bar{e}(t)\|$, inter-event intervals are plotted in Figure 8.3(c) and Figure 8.3(d), respectively.

Next, we will show the effectiveness of the event-triggered control approach for the discrete UIS case. By using the same gain matrices K and L, and assuming the sampling period of the UIS is $T_s = 0.04s$ and $\sigma = 0.5$, it can be verified that the conditions (8.17)–(8.19) are feasible. Then, with the same parameters ε, α, and ε_0, under the event triggering condition (8.16), the simulation results are shown in Figure 8.4, which illustrates the effectiveness of the event-triggered control approach.

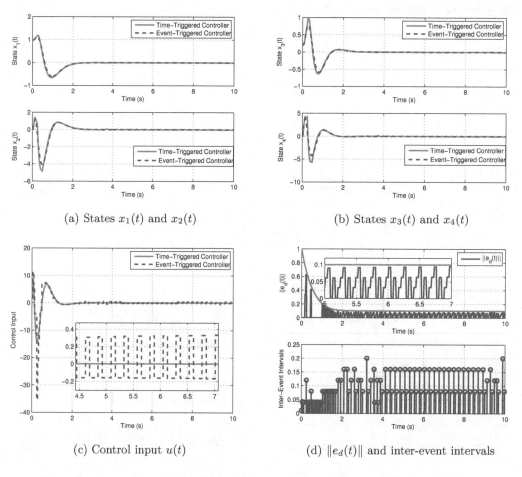

(a) States $x_1(t)$ and $x_2(t)$ (b) States $x_3(t)$ and $x_4(t)$

(c) Control input $u(t)$ (d) $\|e_d(t)\|$ and inter-event intervals

Figure 8.4 Comparative simulation results of time- and event-triggered control approaches (discrete UIS case)

8.6 SUMMARY

For both continuous and discrete UIS cases, the problem of observer-based event-triggered output feedback control of linear systems has been investigated in this chapter. Different from sampled-data control systems, the controller is only updated only when some error signals exceeding a given threshold. It has been shown that the global uniform ultimate boundedness stability of the closed-loop system can be established. A numerical example has been exploited to illustrate the effectiveness of the developed results.

CHAPTER 9

Observer-Based Self-Triggered Control for CPSs

9.1 INTRODUCTION

In Chapter 8, the event-triggered output feedback control problem has been addressed for continuous-time linear systems. For discrete-time systems, some event-triggered control techniques are presented in [26], which are also implemented in the model predictive controller. In [92], event-triggered finite-horizon output-feedback problems of discrete-time linear systems are considered, and a computationally tractable approach is presented to determine a suboptimal event triggering condition. In [52], both global and local event-triggered control approaches are proposed for discrete-time systems. It should be pointed output that, although the conventional event-triggered control strategies can reduce the communication frequency and the resource usage, the continuous monitoring of the plant is required permanently. To overcome this shortcomings of continuous event-triggered control, one approach is the event-triggered control with discrete UIS considered in Chapter 8, the other approach is the so-called self-triggered control strategies [157, 108, 4, 2, 28], where the continuous monitoring of the triggering condition is not required. In [157], the self-triggered control scheme with guaranteed L_2 stability is proposed. In [108], the self-triggered implementation of linear controllers are proposed to reduce the amount of controller updates which are necessary to retain the exponential input-to-state stability of the closed-loop system with respect to additive disturbances. In [4], by using the current state of the plant to decide the next updating instant of the controller, the state feedback self-triggered control approaches are developed for state-dependent homogeneous systems and polynomial systems. In [2], the output feedback self-triggered control strategy is proposed for linear plants with unknown disturbances.

Different from [3, 4], where the state- and output-based self-triggered controllers are proposed for continuous-time linear systems, in this chapter, we focus on the self-triggered output feedback control problem for discrete-time systems, and the UIS is implemented to determine when the controller is updated. For both the full- and reduced-order observer cases, the updating instants are determined, respectively, where only the information of the estimated state at the current updating instant

DOI: 10.1201/9781003260882-9

is required to obtain the next updating instant. Simulation results are presented to show that under the proposed self-triggered control schemes, the updating frequency of the controller is significantly reduced with guaranteed control performances.

9.2 SYSTEM DESCRIPTION

As shown in Figure 9.1, in this chapter, we will study the self-triggered output feedback control problems of discrete-time linear systems.

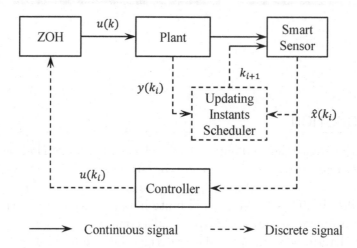

Figure 9.1 Block diagram of self-triggered control system

The discrete-time plant is described by

$$
\begin{aligned}
x(k + 1) &= Ax(k) + Bu(k) \\
y(k) &= Cx(k),
\end{aligned}
\tag{9.1}
$$

where $x(k) \in \mathbb{R}^n$ is the system state, $u(k) \in \mathbb{R}^m$ is the control input, and $y(k) \in \mathbb{R}^q$ is the system output. It is naturally assumed that the pairs (A, B) and (C, A) are stabilizable and detectable, respectively.

Here, we assume that the states are not available for designing self-triggered conditions or controllers, and that the sensor possesses the necessary computation capability and it can pre-process measurement $y(k)$ to obtain the state estimate $\hat{x}(k)$ according to the state observer. For simplicity, the inter-event interval is defined as $[k_i, k_{i+1}) \triangleq \{k_i, k_i + 1, \ldots, k_{i+1} - 1\}$. Then, the controller is given by the following form

$$
u(k) = u(k_i), \ k \in [k_i, k_{i+1}),
\tag{9.2}
$$

where k_i, $i \in \mathbb{N} \triangleq \{0, 1, 2, 3, \ldots\}$ are some integers with $\{k_0, k_1, k_2, k_3, \ldots\} \subset \mathbb{N}$, and k_i, $i = 0, 1, 2, \ldots$, are referred to as updating instants determined by a predefined

UIS, and it is obvious that $k_{i+1} > k_i$. Note that ZOH is implemented to hold the control input until the next event happens at k_{i+1}, i.e., $u(k)$ remains constant in the inter-event interval $[k_i, k_{i+1})$.

In this chapter, the observer-based self-triggered output feedback control problem of discrete-time systems is considered. More specifically, we will design the UISs to determine the updating instant k_{i+1} by using only the information of the estimated state at the current updating instant k_i. Then, with the defined updating instants, the stability of the closed-loop systems is analyzed for both the full- and reduced-order observer cases.

9.3 FULL-ORDER OBSERVER-BASED SELF-TRIGGERED CONTROL

In this section, the following full-order state observer is incorporated in the smart sensor to estimate the unknown state $x(k)$:

$$\hat{x}(k+1) = A\hat{x}(k) + Bu(k) + L(y(k) - C\hat{x}(k)), \tag{9.3}$$

where $\hat{x}(k) \in \mathbb{R}^n$ is the observer state, and L is the observer gain with appropriate dimension which can be designed by the modern control approaches, such as the pole assignment method. Based on the state estimate $\hat{x}(k)$, controller (9.2) can be defined by

$$u(k) = K\hat{x}(k_i), \ k \in [k_i, k_{i+1}), \tag{9.4}$$

where K is the controller gain with appropriate dimension, and k_i, $i = 0, 1, 2, \ldots$, denote the updating instants of the controller and will be determined later. Without loss of generality, we assume that $k_0 = 0$ and the first event occurs at k_0.

Remark 9.1. *It is obvious that, at updating instant k_i, the controller will receive the newest estimated state $\hat{x}(k_i)$, which is held until the next event occurs. In this situation, the updating frequency of the controller can be reduced.*

Define $\tilde{x}(k) = x(k) - \hat{x}(k)$ and $e(k) = \hat{x}(k) - \hat{x}(k_i)$. For $k \in [k_i, k_{i+1})$, system (9.1) with controller (9.4) can be reorganized as

$$
\begin{aligned}
x(k+1) &= Ax(k) + Bu(k) \\
&= Ax(k) + BK\hat{x}(k_i) + BK\hat{x}(k) - BK\hat{x}(k) + BKx(k) - BKx(k) \\
&= (A + BK)x(k) - BK\tilde{x}(k) - BKe(k), \tag{9.5}
\end{aligned}
$$

and

$$\tilde{x}(k+1) = x(k+1) - \hat{x}(k+1) = (A - LC)\tilde{x}(k). \tag{9.6}$$

By introducing an augmented variable $\xi(k) = [x^T(k), \tilde{x}^T(k)]^T$, we have the following closed-loop system

$$\xi(k+1) = \bar{A}\xi(k) + \bar{B}e(k), \ \ k \in [k_i, k_{i+1}), \tag{9.7}$$

where $\bar{A} = \begin{bmatrix} A + BK & -BK \\ 0 & A - LC \end{bmatrix}$, $\bar{B} = \begin{bmatrix} -BK \\ 0 \end{bmatrix}$. The next theorem presents the definition of the updating instants, and shows that the states of the closed-loop system (9.7) are globally UUB.

Theorem 9.1. *Suppose that there exists a scalar $\beta > 0$ satisfying that $\|x(0) - \hat{x}(0)\| \leq \beta$, where the $\hat{x}(0)$ is the estimate of the initial state $x(0)$. Consider the closed-loop system (9.7) with updating instants determined by*

$$k_{i+1} = \sup\left\{ k > k_i \Big| \|M^{k-k_i}\| \|(A - LC)^{k_i}\| \beta + \left\| \sum_{j=0}^{k-k_i-1} M^j N \hat{x}(k_i) \right\| \leq \sqrt{\varepsilon} \right\}, \quad \forall i \in \mathbb{N},$$

$$(9.8)$$

where $\varepsilon > 0$ is an event threshold that is appropriately chosen in advance, and

$$M = \begin{bmatrix} A & LC \\ 0 & A - LC \end{bmatrix}, \quad N = \begin{bmatrix} A + BK - I \\ 0 \end{bmatrix}. \tag{9.9}$$

If there exists a positive definite matrix P satisfying the following matrix inequality

$$Q \triangleq \bar{A}^T P \bar{A} - P + \bar{A}^T P \bar{B} \bar{B}^T P \bar{A} < 0, \tag{9.10}$$

then the states of the closed-loop system (9.7) are globally UUB.

Proof. The proof is twofold, we first determine the updating instants (9.8), then we show the uniform ultimate boundedness of the states of the closed-loop system (9.7). It is easy to see that

$$\begin{aligned} e(k+1) &= \hat{x}(k+1) - \hat{x}(k_i) \\ &= A\hat{x}(k) + BK\hat{x}(k_i) + Ly(k) - LC\hat{x}(k) - \hat{x}(k_i) \\ &= A\hat{x}(k) + LC\tilde{x}(k) + (BK - I)\hat{x}(k_i) \\ &= Ae(k) + LC\tilde{x}(k) + (A + BK - I)\hat{x}(k_i), \quad k \in [k_i, k_{i+1}). \end{aligned}$$

Define an augmented state variable $\omega(k) = [e^T(k), \tilde{x}^T(k)]^T$ and thus $\omega(k_i) = [0, \tilde{x}^T(k_i)]^T$. Using (9.6) and (9.9), we have

$$\omega(k+1) = M\omega(k) + N\hat{x}(k_i), \quad k \in [k_i, k_{i+1}).$$

Then, it is straightforward to arrive at

$$\omega(k) = M^{k-k_i}\omega(k_i) + \sum_{j=0}^{k-k_i-1} M^j N \hat{x}(k_i), \quad k = k_i + 1, \ldots, k_{i+1} - 1,$$

which further gives

$$e(k) = \begin{bmatrix} I_n & 0 \end{bmatrix} \omega(k) = \begin{bmatrix} I_n & 0 \end{bmatrix} \left(M^{k-k_i}\omega(k_i) + \sum_{j=0}^{k-k_i-1} M^j N \hat{x}(k_i) \right),$$

$$k = k_i + 1, \ldots, k_{i+1} - 1. \tag{9.11}$$

Noting that $\omega(k_i) = [0, \tilde{x}^T(k_i)]^T$, we have $\|\omega(k_i)\| = \|\tilde{x}(k_i)\|$. It follows from (9.6) that $\tilde{x}(k_i) = (A - LC)^{k_i}\tilde{x}(0)$, which together with $\|\tilde{x}(0)\| \leq \beta$ implies that

$$\|e(k)\| = \left\| \begin{bmatrix} I_n & 0 \end{bmatrix} \left(M^{k-k_i}\omega(k_i) + \sum_{j=0}^{k-k_i-1} M^j N\hat{x}(k_i) \right) \right\|$$

$$\leq \left\| \begin{bmatrix} I_n & 0 \end{bmatrix} \right\| \left(\|M^{k-k_i}\| \|\omega(k_i)\| + \left\| \sum_{j=0}^{k-k_i-1} M^j N\hat{x}(k_i) \right\| \right)$$

$$= \|M^{k-k_i}\| \|\tilde{x}(k_i)\| + \left\| \sum_{j=0}^{k-k_i-1} M^j N\hat{x}(k_i) \right\|$$

$$= \|M^{k-k_i}\| \|(A - LC)^{k_i}\tilde{x}(0)\| + \left\| \sum_{j=0}^{k-k_i-1} M^j N\hat{x}(k_i) \right\|$$

$$\leq \|M^{k-k_i}\| \|(A - LC)^{k_i}\|\beta + \left\| \sum_{j=0}^{k-k_i-1} M^j N\hat{x}(k_i) \right\|.$$

Note that $\| \begin{bmatrix} I_n & 0 \end{bmatrix} \|_2 = \sigma_{\max}(\begin{bmatrix} I_n & 0 \end{bmatrix}) = 1$, thus the updating instants can be defined by (9.8), which indicates that only the information of state $\hat{x}(k_i)$ is required to determine the next updating instant k_{i+1}.

Next, we will show the uniform ultimate boundedness of the closed-loop system (9.7) with updating instants (9.8). To get the desired results, we introduce the Lyapunov function $V(k) = \xi^T(k)P\xi(k)$ with $P > 0$ being a solution of (9.10). For $k \in [k_i, k_{i+1})$, it can be verified that

$$\begin{aligned} \Delta V(k) &= V(k+1) - V(k) \\ &= \xi^T(k+1)P\xi(k+1) - \xi^T(k)P\xi(k) \\ &= \xi^T(k)(\bar{A}^T P\bar{A} - P)\xi(k) + \xi^T(k)\bar{A}^T P\bar{B}e(k) \\ &\quad + e^T(k)\bar{B}^T P\bar{A}\xi(k) + e^T(k)\bar{B}^T P\bar{B}e(k) \\ &= \xi^T(k)(\bar{A}^T P\bar{A} - P + \bar{A}^T P\bar{B}\bar{B}^T P\bar{A})\xi(k) \\ &\quad - \left\| e(k) - \bar{B}^T P\bar{A}\xi(k) \right\|^2 + \|e(k)\|^2 + e^T(k)\bar{B}^T P\bar{B}e(k) \\ &\leq -\lambda_{\min}(-Q)\|\xi(k)\|^2 + (1 + \|\bar{B}^T P\bar{B}\|)\|e(k)\|^2 \\ &\leq -\frac{\lambda_{\min}(-Q)}{\lambda_{\max}(P)}\xi^T(k)P\xi(k) + (1 + \|\bar{B}^T P\bar{B}\|)\|e(k)\|^2 \\ &= -\delta V(k) + \mu\|e(k)\|^2, \end{aligned}$$

where $\delta = \frac{\lambda_{\min}(-Q)}{\lambda_{\max}(P)} < 1$, $\mu = 1 + \|\bar{B}^T P\bar{B}\|$. According to the definition of the updating instants (9.8), we have $\|e(k)\| \leq \sqrt{\varepsilon}$ for $k \in [k_i, k_{i+1})$, which further gives

$$\Delta V(k) \leq -\delta V(k) + c, \quad \forall k \in [k_i, k_{i+1}) \tag{9.12}$$

with $c = \mu\varepsilon$.

Now, we are in a position to show that the states of the closed-loop system (9.7) with updating instants (9.8) are globally UUB.

According to (9.12), it is easy to show that the following inequality holds for $k = k_i, k_i + 1, \ldots, k_{i+1} - 2$,

$$V(k) \le (1-\delta)^{k-k_i} V(k_i) + \sum_{j=0}^{k-k_i-1} (1-\delta)^j c. \qquad (9.13)$$

When $k = k_{i+1} - 1$, it follows from (9.12) that $V(k_{i+1}) \le (1-\delta)V(k_{i+1} - 1) + c$. Thus, by iterations, we have

$$
\begin{aligned}
V(k_{i+1}) &\le (1-\delta)V(k_{i+1} - 1) + c \\
&\le (1-\delta)^2 V(k_{i+1} - 2) + (1-\delta)c + c \\
&\ \ \vdots \\
&\le (1-\delta)^{k_{i+1}-k_i} V(k_i) + \sum_{j=0}^{k_{i+1}-k_i-1} (1-\delta)^j c. \qquad (9.14)
\end{aligned}
$$

Therefore, it can be concluded that inequality (9.13) holds for $k \in [k_i, k_{i+1})$. Then, it follows from (9.13) and (9.14) that

$$
\begin{aligned}
V(k) &\le (1-\delta)^{k-k_i} \left((1-\delta)^{k_i-k_{i-1}} V(k_{i-1}) + \sum_{j=0}^{k_i-k_{i-1}-1} (1-\delta)^j c \right) + \sum_{j=0}^{k-k_i-1} (1-\delta)^j c \\
&= (1-\delta)^{k-k_{i-1}} V(k_{i-1}) + \sum_{j=0}^{k_i-k_{i-1}-1} (1-\delta)^{k-k_i+j} c + \sum_{j=0}^{k-k_i-1} (1-\delta)^j c \\
&= (1-\delta)^{k-k_{i-1}} V(k_{i-1}) + \sum_{j=0}^{k-k_{i-1}-1} (1-\delta)^j c \\
&\ \ \vdots \\
&\le (1-\delta)^{k-k_0} V(k_0) + \sum_{j=0}^{k-k_0-1} (1-\delta)^j c \\
&\le (1-\delta)^{k-k_0} V(k_0) + \frac{c}{\delta},
\end{aligned}
$$

where the last inequality uses the fact that $\sum_{j=0}^{k-k_0-1}(1-\delta)^j c \le \sum_{j=0}^{\infty}(1-\delta)^j c = \frac{c}{\delta}$.

Based on the above deductions, we get

$$
\begin{aligned}
\lambda_{\min}(P)\|\xi(k)\|^2 &\le V(k) \le (1-\delta)^{k-k_0} V(k_0) + \frac{c}{\delta} \\
&\le (1-\delta)^{k-k_0} \lambda_{\max}(P)\|\xi(k_0)\|^2 + \frac{c}{\delta},
\end{aligned}
$$

which yields $\|\xi(k)\|^2 \le (1-\delta)^{k-k_0} \frac{\lambda_{\max}(P)}{\lambda_{\min}(P)} \|\xi(k_0)\|^2 + \frac{c}{\delta \lambda_{\min}(P)}$. This implies that the states of the closed-loop system (9.7) exponentially converge to the bounded region

$\mathcal{B} = \left\{ \xi(k) \middle| \|\xi(k)\| \leq \sqrt{\frac{c}{\delta \lambda_{\min}(P)}} \right\}$, thus the global uniform ultimate boundedness of the resulting closed-loop system (9.7) can be guaranteed. □

Remark 9.2. *It can be seen from (9.8) that to compute the next updating instant, only the current estimated state is required, which means that, at the current updating instant, it is already known when the next updating will take place.*

Remark 9.3. *It is worth mentioning that, if there exists disturbance in system (9.1), that is,*

$$x(k+1) = Ax(k) + Bu(k) + Ed(k),$$
$$y(k) = Cx(k),$$

where the disturbance model is $d(k+1) = Sd(k)$ with S being a matrix with appropriate dimension [67], then the observer (9.3) can be written in the following extended form as

$$\begin{bmatrix} \hat{x}(k+1) \\ \hat{d}(k+1) \end{bmatrix} = \begin{bmatrix} A & E \\ 0 & S \end{bmatrix} \begin{bmatrix} \hat{x}(k) \\ \hat{d}(k) \end{bmatrix} + \begin{bmatrix} B \\ 0 \end{bmatrix} u(k) + L(y(k) - C\hat{x}(k)).$$

Based on the state estimate $\hat{x}(t)$, the controller can be defined by $u(k) = K\hat{x}(k_i)$. Thus, the proposed self-triggered control approach can be extended to this general case directly.

9.4 REDUCED-ORDER OBSERVER-BASED SELF-TRIGGERED CONTROL

In this section, we consider the reduced-order observer-based self-triggered output feedback control problem for system (9.1).

Here we assume that part of the state $x(k)$ can be obtained from $y(k)$ directly. Without loss of generality, assume $C = [I_q, \ 0]$, and partitions A, B in (9.1) respectively as $A = \begin{bmatrix} A_{11} & A_{12} \\ A_{21} & A_{22} \end{bmatrix}$, $B = \begin{bmatrix} B_1 \\ B_2 \end{bmatrix}$, where $A_{11} \in \mathbb{R}^{q \times q}$, $A_{12} \in \mathbb{R}^{q \times (n-q)}$, $A_{21} \in \mathbb{R}^{(n-q) \times q}$, $A_{22} \in \mathbb{R}^{(n-q) \times (n-q)}$, $B_1 \in \mathbb{R}^{q \times m}$, $B_2 \in \mathbb{R}^{(n-q) \times m}$.

Let $z(k) = [z_1^T(k), z_2^T(k)]^T$. For system (9.1), we take a nonsingular state transformation $z(k) = Tx(k)$, where $T = \begin{bmatrix} I_q & 0 \\ L_r & I_{n-q} \end{bmatrix}$. Then, applying this state transformation onto (9.1) gives

$$z(k+1) = \hat{A}z(k) + \hat{B}u(k),$$
$$y(k) = z_1(k), \tag{9.15}$$

where $\hat{A} = \begin{bmatrix} \hat{A}_{11} & \hat{A}_{12} \\ \hat{A}_{21} & \hat{A}_{22} \end{bmatrix}$, $\hat{B} = \begin{bmatrix} \hat{B}_1 \\ \hat{B}_2 \end{bmatrix}$, with $\hat{A}_{11} = A_{11} - A_{12}L_r$, $\hat{A}_{12} = A_{12}$, $\hat{A}_{21} = L_r A_{11} + A_{21} - L_r A_{12} L_r - A_{22} L_r$, $\hat{A}_{22} = L_r A_{12} + A_{22}$, $\hat{B}_1 = B_1$, $\hat{B}_2 = L_r B_1 + B_2$, respectively, and L_r is selected such that $A_{22} + L_r A_{12}$ is Schur.

Lemma 9.1. *If the pair (C, A) is detectable, then the pair (A_{22}, A_{12}) is detectable, and there exists the reduced-order observer gain L_r such that $A_{22} + L_r A_{12}$ is Schur.*

Proof. Since the pair (C, A) is detectable, we have

$$
\mathrm{rank} \begin{bmatrix} \lambda I_n - A \\ C \end{bmatrix} = \mathrm{rank} \begin{bmatrix} \lambda I_q - A_{11} & -A_{12} \\ -A_{21} & \lambda I_{n-q} - A_{22} \\ I_q & 0 \end{bmatrix} = n, \quad \forall \lambda \in \mathbb{C}, \quad |\lambda| \geq 1,
$$

which indicates that matrix $\begin{bmatrix} -A_{12} \\ \lambda I_{n-q} - A_{22} \\ 0 \end{bmatrix}$ is also of full column rank, i.e.,

$$
\mathrm{rank} \begin{bmatrix} -A_{12} \\ \lambda I_{n-q} - A_{22} \\ 0 \end{bmatrix} = \mathrm{rank} \begin{bmatrix} -A_{12} \\ \lambda I_{n-q} - A_{22} \end{bmatrix} = n - q, \quad \forall \lambda \in \mathbb{C}, \quad |\lambda| \geq 1.
$$

Thus, we can conclude that the pair (A_{22}, A_{12}) is detectable, and there exists gain matrix L_r such that $A_{22} + L_r A_{12}$ is Schur. This completes the proof. □

From (9.15), we can obtain

$$
z_2(k + 1) = \hat{A}_{22} z_2(k) + \hat{B}_2 u(k) + \hat{A}_{21} y(k),
$$

so the reduced-order observer can be designed as follows

$$
\hat{z}_2(k + 1) = \hat{A}_{22} \hat{z}_2(k) + \hat{B}_2 u(k) + \hat{A}_{21} y(k). \tag{9.16}
$$

Then, the reduced-order observer-based controller for system (9.15) is given by

$$
u(k) = K_r \hat{z}(k_i), \quad k \in [k_i, k_{i+1}), \tag{9.17}
$$

where $\hat{z}(k_i) = \begin{bmatrix} \hat{z}_1(k_i) \\ \hat{z}_2(k_i) \end{bmatrix} = \begin{bmatrix} y(k_i) \\ \hat{z}_2(k_i) \end{bmatrix}$.

Define the estimation error $\tilde{z}(k) = z(k) - \hat{z}(k)$. Noting that $z_1(k) = \hat{z}_1(k) = y(k)$, we have

$$
\bar{e}_r(k) = \hat{z}(k) - \hat{z}(k_i) = \begin{bmatrix} y(k) - y(k_i) \\ \hat{z}_2(k) - \hat{z}_2(k_i) \end{bmatrix}, \quad k \in [k_i, k_{i+1}).
$$

Thus, the closed-loop system can be expressed as

$$
\begin{aligned}
z(k + 1) &= \hat{A} z(k) + \hat{B} K_r \hat{z}(k_i) \\
&= \hat{A} z(k) + \hat{B} K_r z(k) - \hat{B} K_r z(k) + \hat{B} K_r \hat{z}(k) - \hat{B} K_r \hat{z}(k) + \hat{B} K_r \hat{z}(k_i) \\
&= (\hat{A} + \hat{B} K_r) z(k) - \hat{B} K_r \tilde{z}(k) - \hat{B} K_r \bar{e}_r(k).
\end{aligned}
$$

It is easy to obtain that $\tilde{z}(k) = \begin{bmatrix} 0 \\ I_{n-q} \end{bmatrix} \tilde{z}_2(k)$, and

$$
\tilde{z}_2(k + 1) = z_2(k + 1) - \hat{z}_2(k + 1) = \hat{A}_{22} \tilde{z}_2(k). \tag{9.18}
$$

Define an augmented variable $\xi_r(k) = [z^T(k), \tilde{z}_2^T(k)]^T$, we have the following closed-loop system

$$\xi_r(k+1) = \bar{A}_r \xi_r(k) + \bar{B}_r \bar{e}_r(k), \quad k \in [k_i, k_{i+1}), \tag{9.19}$$

where

$$\bar{A}_r = \begin{bmatrix} \hat{A} + \hat{B}K_r & -\hat{B}K_r \begin{bmatrix} 0 \\ I_{n-q} \end{bmatrix} \\ 0 & \hat{A}_{22} \end{bmatrix}, \quad \bar{B}_r = \begin{bmatrix} -\hat{B}K_r \\ 0 \end{bmatrix}.$$

Now, we consider the stability analysis problem of the closed-loop system (9.19). The following theorem gives the definition of the updating instants and shows that the states of the closed-loop system (9.19) are globally UUB.

Theorem 9.2. *Let $\hat{z}_2(0)$ be an estimate of the initial condition $z_2(0)$, and assume that there exists a scalar $\varpi > 0$ such that $\|z_2(0) - \hat{z}_2(0)\| \leq \varpi$. Consider the closed-loop system (9.19) with updating instants determined by*

$$k_{i+1} = \sup\left\{k > k_i \,\middle|\, \|H^{k-k_i}\|\|\hat{A}_{22}^{k_i}\|\varpi + \left\|\sum_{j=0}^{k-k_i-1} H^j G\hat{z}(k_i)\right\| \leq \sqrt{\varepsilon}\right\}, \quad \forall i \in \mathbb{N}, \tag{9.20}$$

where $\varepsilon > 0$ is an event threshold that is appropriately chosen in advance, and

$$H = \begin{bmatrix} \hat{A}_{11} & \hat{A}_{12} & \hat{A}_{12} \\ \hat{A}_{21} & \hat{A}_{22} & 0 \\ 0 & 0 & \hat{A}_{22} \end{bmatrix}, \quad G = \begin{bmatrix} \hat{B}_1 K_r + \begin{bmatrix} \hat{A}_{11} - I_q & \hat{A}_{12} \end{bmatrix} \\ \hat{B}_2 K_r + \begin{bmatrix} \hat{A}_{21} & \hat{A}_{22} - I_{n-q} \end{bmatrix} \\ 0 \end{bmatrix}. \tag{9.21}$$

If there exists a positive definite matrix P satisfying the following matrix inequality

$$Q_r \triangleq \bar{A}_r^T P \bar{A}_r - P + \bar{A}_r^T P \bar{B}_r \bar{B}_r^T P \bar{A}_r < 0, \tag{9.22}$$

then the states of the closed-loop system (9.19) are globally UUB.

Proof. Since $\hat{z}(k_i)$ is constant in the inter-event interval $[k_i, k_{i+1})$, it can be deduced from (9.19) that

$$\begin{aligned}
e_r(k+1) &= y(k+1) - y(k_i) \\
&= z_1(k+1) - z_1(k_i) \\
&= \hat{A}_{11} z_1(k) + \hat{A}_{12} z_2(k) + \hat{B}_1 u(k) - z_1(k_i) \\
&= \hat{A}_{11} \hat{z}_1(k) + \hat{A}_{12} z_2(k) + \hat{B}_1 K_r \hat{z}(k_i) - \hat{z}_1(k_i) \\
&= \begin{bmatrix} \hat{A}_{11} & \hat{A}_{12} \end{bmatrix} \bar{e}_r(k) + \hat{A}_{12} \tilde{z}_2(k) \\
&\quad + \left(\hat{B}_1 K_r + \begin{bmatrix} \hat{A}_{11} & \hat{A}_{12} \end{bmatrix} - \begin{bmatrix} I_q & 0 \end{bmatrix}\right) \hat{z}(k_i),
\end{aligned}$$

and

$$
\begin{aligned}
\hat{e}(k+1) &= \hat{z}_2(k+1) - \hat{z}_2(k_i) \\
&= \hat{A}_{22}\hat{z}_2(k) + \hat{B}_2 u(k) + \hat{A}_{21}y(k) - \hat{z}_2(k_i) \\
&= \hat{A}_{22}\hat{z}_2(k) + \hat{B}_2 K_r \hat{z}(k_i) + \hat{A}_{21}\hat{z}_1(k) - \hat{z}_2(k_i) \\
&= \begin{bmatrix} \hat{A}_{21} & \hat{A}_{22} \end{bmatrix} \bar{e}_r(k) + \left(\hat{B}_2 K_r + \begin{bmatrix} \hat{A}_{21} & \hat{A}_{22} \end{bmatrix} - \begin{bmatrix} 0 & I_{n-q} \end{bmatrix} \right) \hat{z}(k_i).
\end{aligned}
$$

Thus, we have the following compact form for the error signal

$$
\begin{aligned}
\bar{e}_r(k+1) &= \begin{bmatrix} \hat{A}_{11} & \hat{A}_{12} \\ \hat{A}_{21} & \hat{A}_{22} \end{bmatrix} \bar{e}_r(k) + \begin{bmatrix} \hat{A}_{12} \\ 0 \end{bmatrix} \tilde{z}_2(k) \\
&\quad + \begin{bmatrix} \hat{B}_1 K_r + \begin{bmatrix} \hat{A}_{11} - I_q & \hat{A}_{12} \end{bmatrix} \\ \hat{B}_2 K_r + \begin{bmatrix} \hat{A}_{21} & \hat{A}_{22} - I_{n-q} \end{bmatrix} \end{bmatrix} \hat{z}(k_i).
\end{aligned}
$$

Define an augmented state variable $\psi(k) = [\bar{e}_r^T(k), \tilde{z}_2^T(k)]^T$. Using (9.18) and (9.21), we obtain

$$
\psi(k+1) = H\psi(k) + G\hat{z}(k_i),
$$

which implies that $\psi(k) = H^{k-k_i}\psi(k_i) + \sum_{j=0}^{k-k_i-1} H^j G\hat{z}(k_i)$. Then we get

$$
\begin{aligned}
\bar{e}_r(k) &= \begin{bmatrix} I_n & 0 \end{bmatrix} \psi(k) \\
&= \begin{bmatrix} I_n & 0 \end{bmatrix} \left(H^{k-k_i}\psi(k_i) + \sum_{j=0}^{k-k_i-1} H^j G\hat{z}(k_i) \right).
\end{aligned} \tag{9.23}
$$

Let the estimation error of the initial condition be $\tilde{z}_2(0) = z_2(0) - \hat{z}_2(0)$. As $\|\tilde{z}_2(0)\| \le \varpi$, it follows from (9.18) that $\tilde{z}_2(k_i) = \hat{A}_{22}^{k_i}\tilde{z}_2(0)$.

Since $\psi(k_i) = [0, \tilde{z}_2^T(k)]^T$, we have $\|\psi(k_i)\| = \|\tilde{z}_2(k)\|$. Then, it follows from (9.23) that

$$
\begin{aligned}
\|\bar{e}_r(k)\| &= \left\| \begin{bmatrix} I_n & 0 \end{bmatrix} \left(H^{k-k_i}\psi(k_i) + \sum_{j=0}^{k-k_i-1} H^j G\hat{z}(k_i) \right) \right\| \\
&\le \left\| \begin{bmatrix} I_n & 0 \end{bmatrix} \right\| \left(\|H^{k-k_i}\| \|\psi(k_i)\| + \left\| \sum_{j=0}^{k-k_i-1} H^j G\hat{z}(k_i) \right\| \right) \\
&= \|H^{k-k_i}\| \|\tilde{z}_2(k_i)\| + \left\| \sum_{j=0}^{k-k_i-1} H^j G\hat{z}(k_i) \right\| \\
&= \|H^{k-k_i}\| \|\hat{A}_{22}^{k_i}\tilde{z}_2(0)\| + \left\| \sum_{j=0}^{k-k_i-1} H^j G\hat{z}(k_i) \right\| \\
&\le \|H^{k-k_i}\| \|\hat{A}_{22}^{k_i}\| \varpi + \left\| \sum_{j=0}^{k-k_i-1} H^j G\hat{z}(k_i) \right\|.
\end{aligned}
$$

Hence, the updating instants can be decided by (9.20), which indicates that only the estimated state $\hat{z}(k_i)$ is needed to determine the next updating instant k_{i+1}. Similar to the proof of Theorem 9.1, it can be shown that the states of the closed-loop system (9.19) are UUB, thus the details are omitted here for brevity. □

9.5 NUMERICAL SIMULATION

Consider the cart with an inverted pendulum system borrowed from Chapter 8. Assume that $M_0 = 1.096kg$, $m_0 = 0.109kg$, $b = 0.1N/m/sec$, $g = 9.8m/s^2$, $l = 0.25m$, and $I = 0.0034kg.m^2$. Taking the sampling period $T_s = 10ms$ yields the following discretized system model

$$x(k+1) = \begin{bmatrix} 1.0000 & 0.0100 & 0.0000 & 0.0000 \\ 0 & 0.9991 & 0.0063 & 0.0000 \\ 0 & -0.0000 & 1.0014 & 0.0100 \\ 0 & -0.0024 & 0.2784 & 1.0014 \end{bmatrix} x(k) + \begin{bmatrix} 0.0000 \\ 0.0088 \\ 0.0001 \\ 0.0236 \end{bmatrix} u(k),$$

$$y(k) = \begin{bmatrix} 1 & 0 & 0 & 0 \\ 0 & 0 & 1 & 0 \end{bmatrix} x(k).$$

In the simulation, the pole assignment technique is used to obtain the controller and observer gains.

We first consider the full-order observer-based self-triggered control case. By choosing the controller poles as 0.95, 0.87, 0.8, 0.96 and the observer poles as 0.4, 0.6, 0.5, 0.65, the feedback gains K and L can be computed respectively as

$$K = \begin{bmatrix} 225.2118 & 126.6515 & -328.5851 & -64.1266 \end{bmatrix},$$

$$L = \begin{bmatrix} 0.9990 & -0.0038 \\ 23.9171 & -0.1565 \\ -0.0071 & 0.8529 \\ -0.4124 & 17.8947 \end{bmatrix}.$$

According to the UIS (9.8) with $\varepsilon = 0.0001$ and $\beta = 0.2646$, the simulation results are plotted in Figure 9.2, where the initial condition is selected as $x(0) = [0, 0, 0.1, 0]^T$. It can be seen that the states of the closed-loop system are globally UUB, and the performance of the self-triggered control system is very close to the time-triggered control system. For the time-triggered control system, there are 700 updating instants of the controller, while only 159 updating instants of the controller are required for the proposed full-order observer-based self-triggered control system. It is obvious to see that the proposed self-triggered control scheme can reduce the updating frequency of the controller significantly, thus lessening the burden of the controller. The inter-event interval is shown in Figure 9.2(d), from which it can be seen that for $k \leq 42$, $e(k) \equiv 0$, which indicates that the updating of the controller happens at each time step. Moreover, the error signal $\|e(k)\|$ is also plotted in Figure 9.2(d). It is seen that $\|e(k)\|$ is less than the given threshold, which agrees with the definition of the updating instants in (9.8).

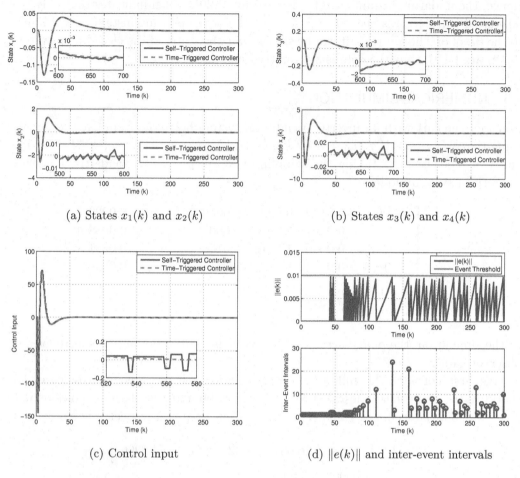

(a) States $x_1(k)$ and $x_2(k)$

(b) States $x_3(k)$ and $x_4(k)$

(c) Control input

(d) $\|e(k)\|$ and inter-event intervals

Figure 9.2 Comparative simulation results of time- and self-triggered control approaches (full-order observer case)

Next, we examine the reduced-order observer case. Taking an equivalent transformation of the discretized pendulum system produces that

$$\bar{x}(k+1) = \begin{bmatrix} 1.0000 & 0.0000 & 0.0100 & 0.0000 \\ 0 & 1.0014 & -0.0000 & 0.0100 \\ 0 & 0.0063 & 0.9991 & 0.0000 \\ 0 & 0.2784 & -0.0024 & 1.0014 \end{bmatrix} \bar{x}(k) + \begin{bmatrix} 0.0000 \\ 0.0001 \\ 0.0088 \\ 0.0236 \end{bmatrix} u(k),$$

$$y(k) = \begin{bmatrix} 1 & 0 & 0 & 0 \\ 0 & 1 & 0 & 0 \end{bmatrix} \bar{x}(k).$$

Then, by choosing the controller poles as 0.86, 0.97, 0.98, 0.96 and the observer poles as 0.87, 0.75, the controller and the observer gains can be calculated respectively as

$$K_r = \begin{bmatrix} 233.3321 & -475.0445 & 16.6860 & -15.7177 \end{bmatrix},$$

$$L_r = \begin{bmatrix} -12.9174 & -0.0030 \\ 0.2061 & -25.1275 \end{bmatrix}.$$

Under the UIS (9.20) with $\varepsilon = 0.0001$ and $\varpi = 3.1014$, the simulation results are plotted in Figure 9.3.

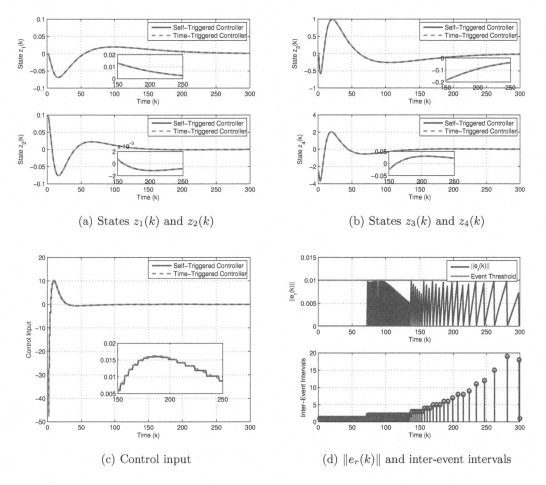

(a) States $z_1(k)$ and $z_2(k)$

(b) States $z_3(k)$ and $z_4(k)$

(c) Control input

(d) $\|e_r(k)\|$ and inter-event intervals

Figure 9.3 Comparative simulation results of time- and self-triggered control approaches (reduced-order observer case)

It is clear that, for the time-triggered control system, there are 700 updating instants of the controller. However, the proposed reduced-order observer-based self-triggered control scheme performs with only 183 updating instants over 700 time-steps. From the simulation results, it is clearly seen that the reduced-order observer-based self-triggered controller also works well.

9.6 SUMMARY

For the discrete-time linear systems, both the full- and reduced-order observer-based self-triggered control problems have been investigated in this chapter, where the updating instants of the controllers have been established by using the estimated states. It has been shown that with the self-triggered control schemes, the states of the corresponding closed-loop systems are UUB. Finally, the applicability and effectiveness of the developed results have been demonstrated by a numerical example.

CHAPTER 10

Event-Triggered Dynamic Output-Feedback Control for CPSs

10.1 INTRODUCTION

In the previous two chapters, the observer-based event-triggered control strategies have been proposed for continuous- and discrete-time systems, respectively. It is worth mentioning that the reconstruction of the states requires to know the current input of the plant. However, in CPSs, because the controller and actuator are distributively located, and there exist time-varying network communication delay and data dropout in the controller-to-actuator channel, it is complicated and challenging to know the current input of the plant at the controller side. Alternatively, output feedback control strategies can be employed in which the control law only requires knowledge of measured output [132, 219, 159, 53, 133]. In [132], a time-delay model-based event-triggered static output feedback control approach is proposed for CPSs with network communication delay. In [219], the event-triggered dynamic output feedback (DOF) controller is designed for CPSs, where the output measurement signals of the physical plant are sampled periodically. In [159], an event-triggered DOF controller with a switching rule is proposed and the global asymptotic stability of the closed-loop systems can be guaranteed. In [53], the event-triggered DOF control problem is addressed for networked nonlinear systems, where an output-based discrete UIS is introduced to choose those only necessary sampled-data packets to be transmitted through a communication network for controller design. In [133], the event-triggered output feedback H_∞ control problem is studied for CPSs, where the interval decomposition method is introduced to place the controlled plant and the output feedback controller into the same updated time interval but with updated signals at different instants.

Based on the above observations, in this chapter, we focus on designing DOF controllers for systems with event-triggered control inputs, which can significantly save the data transmissions and in the meanwhile maintain the system performance at a satisfactory level. The DOF controllers which transmit the control signal based on a predefined UIS, are designed for systems with either continuous or sampled output measurements. It is shown that, with the proposed event-triggered control scheme, the global uniform ultimate boundedness of the closed-loop systems is guaranteed and the inter-event interval is lower bounded by a positive scalar. Finally, a numerical example is used to verify the effectiveness and the merits of the proposed controller design techniques.

10.2 SYSTEM DESCRIPTION

In this chapter, we consider the following linear time-invariant (LTI) plant with event-triggered control inputs,

$$
\begin{aligned}
\dot{x}(t) &= Ax(t) + Bu(t_k), \\
y(t) &= Cx(t), \quad t \in [t_k, t_{k+1}), \quad k \in \mathbb{N},
\end{aligned}
\tag{10.1}
$$

where $x(t) \in \mathbb{R}^n$ represents the state vector, $u(t) \in \mathbb{R}^m$ denotes the control input, $y(t) \in \mathbb{R}^q$ is the measured output, and A, B, C are the system matrices with appropriate dimensions. The controller updating instants $\{t_k\}_{k \in \mathbb{N}}$ are defined as an increasing sequence of positive scalars with $\bigcup_{k \in \mathbb{N}} [t_k, t_{k+1}) = [0, +\infty)$, and $\{t_k\}_{k \in \mathbb{N}}$ are determined by a predefined UIS. To hold the control signal continuous, a ZOH is embedded, thus, during the inter-event interval $[t_k, t_{k+1})$, $u(t) \equiv u(t_k)$.

In this chapter, the DOF control problem of continuous-time systems with event-triggered control inputs is considered. More specifically, we will design the UISs and DOF controllers for systems with both continuous and sampled output measurements, and the analysis of the avoidance of the Zeno behavior and the stability of the closed-loop system are also performed.

10.3 EVENT-TRIGGERED DOF CONTROL FOR SYSTEMS WITH CONTINUOUS OUTPUT MEASUREMENTS

In this section, by assuming that the continuous output measurements $y(t)$ are available, we design the DOF controller as

$$
\begin{aligned}
\dot{\hat{x}}_c(t) &= K_1 \hat{x}_c(t) + K_2 y(t) \\
u(t) &= K_3 \hat{x}_c(t) + K_4 y(t),
\end{aligned}
\tag{10.2}
$$

where the $\hat{x}_c(t) \in \mathbb{R}^n$ is the controller state, and K_i, $i = 1, 2, 3, 4$, denote the controller parameter matrices to be ascertained. Thus, for $t \in [t_k, t_{k+1})$, the plant and controller can be reorganized as:

$$
\begin{aligned}
\dot{x}(t) &= Ax(t) + Bu(t_k) \\
&= Ax(t) + Bu(t) - Bu(t) + Bu(t_k) \\
&= Ax(t) + BK_3 \hat{x}_c(t) + BK_4 y(t) - Be(t) \\
&= (A + BK_4 C)x(t) + BK_3 \hat{x}_c(t) - Be(t),
\end{aligned}
$$

where $e(t) = u(t) - u(t_k)$ and

$$\dot{\hat{x}}_c(t) = K_1 \hat{x}_c(t) + K_2 y(t) = K_2 C x(t) + K_1 \hat{x}_c(t).$$

Defining an augmented state $\xi(t) = [x^T(t), \hat{x}_c^T(t)]^T$, we arrive at the closed-loop system as

$$\dot{\xi}(t) = \bar{A} \xi(t) + \bar{B} e(t), \quad t \in [t_k, t_{k+1}), \tag{10.3}$$

where $\bar{A} = \begin{bmatrix} A + BK_4C & BK_3 \\ K_2C & K_1 \end{bmatrix}$, $\bar{B} = \begin{bmatrix} -B \\ 0 \end{bmatrix}$. Moreover, the stability of closed-loop system (10.3) is shown in the following theorem.

Theorem 10.1. *For the closed-loop system (10.3), the following UIS is designed to determine the updating instants,*

$$t_{k+1} = \sup \left\{ t > t_k | \|e(t)\|^2 \leq \varepsilon + \varrho \|u(t)\|^2 \right\}, \quad \forall k \in \mathbb{N}, \tag{10.4}$$

where $\varepsilon > 0$ and $\varrho > 0$ are two user-defined parameters. If the following matrix inequality

$$Q \triangleq \bar{A}^T P + P\bar{A} + P\bar{B}\bar{B}^T P + \varrho \bar{C}^T \bar{C} < 0, \tag{10.5}$$

holds with $\bar{C} = [K_4C, K_3]$ for a symmetric and positive definite matrix P, then the states of system (10.3) are globally UUB.

Proof. With a symmetric and positive definite matrix $P > 0$ which is the solution of (10.5), the Lyapunov function candidate is chosen as $V(t) = \xi^T(t) P \xi(t)$ for $t \in [t_k, t_{k+1})$. Then, the derivation of $V(t)$ is

$$
\begin{aligned}
\dot{V}(t) &= \dot{\xi}^T(t) P \xi(t) + \xi^T(t) P \dot{\xi}(t) \\
&= (\bar{A}\xi(t) + \bar{B}e(t))^T P \xi(t) + \xi^T(t) P(\bar{A}\xi(t) + \bar{B}e(t)) \\
&= \xi^T(t)(\bar{A}^T P + P\bar{A})\xi(t) + e^T(t)\bar{B}^T P \xi(t) + \xi^T(t) P\bar{B}e(t) \\
&= \xi^T(t)(\bar{A}^T P + P\bar{A} + P\bar{B}\bar{B}^T P)\xi(t) - \|e(t) - \bar{B}^T P \xi(t)\|^2 + \|e(t)\|^2.
\end{aligned}
$$

It can be seen that, with the updating instants defined by (10.4), $\|e(t)\|^2 \leq \varepsilon + \varrho \|u(t)\|^2$ always holds for $t \in [t_k, t_{k+1})$. Hence, for $\forall t \in [t_k, t_{k+1})$, we have

$$
\begin{aligned}
\dot{V}(t) &\leq \xi^T(t)(\bar{A}^T P + P\bar{A} + P\bar{B}\bar{B}^T P)\xi(t) + \|e(t)\|^2 \\
&= \xi^T(t)(\bar{A}^T P + P\bar{A} + P\bar{B}\bar{B}^T P + \varrho \bar{C}^T \bar{C})\xi(t) + \varepsilon \\
&\leq -\frac{\lambda_{\min}(-Q)}{\lambda_{\max}(P)} \xi^T(t) P \xi(t) + \varepsilon \\
&= -\delta V(t) + \varepsilon, \tag{10.6}
\end{aligned}
$$

where $\delta = \frac{\lambda_{\min}(-Q)}{\lambda_{\max}(P)}$. From inequality (10.6), we obtain that the Lyapunov function $V(t)$ will decay in each time interval $[t_k, t_{k+1})$. By the comparison lemma in [202], inequality (10.6) is obtained as $V(t) \leq e^{-\delta t}\left(V(0) - \frac{\varepsilon}{\delta}\right) + \frac{\varepsilon}{\delta}$, which leads to $V(t) \leq e^{-\delta(t-t_k)} V(t_k) + \frac{\varepsilon}{\delta}\left(1 - e^{-\delta(t-t_k)}\right)$, and thus we have

$$\|\xi(t)\| \leq \sqrt{\frac{\lambda_{\max}(P)}{\lambda_{\min}(P)}\|\xi(t_k)\|^2 + \frac{\varepsilon}{\delta\lambda_{\min}(P)}}. \tag{10.7}$$

Therefore, the states of closed-loop system (10.3) are global UUB, which completes the proof. □

For avoidance of the Zeno behavior, in the next, it will be shown that a positive lower bound of the minimum inter-event interval $t_{\min} \triangleq \min_{k \in \mathbb{N}}\{t_{k+1} - t_k\}$ always exists for the proposed event-triggered control strategy.

Theorem 10.2. *For the minimum inter-event interval $t_{\min}$, there always exists a positive lower bound $t_{\min}^l > 0$ satisfying that $t_{\min} \geq t_{\min}^l > 0$.*

Proof. For $\forall k \in \mathbb{N}$, let t_k be an arbitrary updating instant. Note that $u(t_k)$ is constant in the time interval $[t_k, t_{k+1})$. Then, for $t \in [t_k, t_{k+1})$, according to the definition of $e(t)$, we have

$$\begin{aligned}
\dot{e}(t) &= K_3\dot{\hat{x}}_c(t) + K_4C\dot{x}(t) \\
&= K_3K_1\hat{x}_c(t) + K_3K_2Cx(t) + K_4CAx(t) \\
&\quad + K_4CBu(t) - K_4CBu(t_k) + K_4CBu(t_k) \\
&= K_4CBe(t) + (K_3K_2C + K_4CA)x(t) \\
&\quad + K_3K_1\hat{x}_c(t) + K_4CBK_4Cx(t_k) + K_4CBK_3\hat{x}_c(t_k) \\
&= \Sigma_1 e(t) + \Sigma_2 \xi(t) + \Sigma_3 \xi(t_k),
\end{aligned}$$

where $\Sigma_1 = K_4CB, \Sigma_2 = \begin{bmatrix} K_3K_2C + K_4CA & K_3K_1 \end{bmatrix}, \Sigma_3 = \begin{bmatrix} K_4CBK_4C & K_4CBK_3 \end{bmatrix}$. Moreover, we have

$$\begin{aligned}
e(t) &= e^{\Sigma_1(t-t_k)}e(t_k) + \int_{t_k}^{t} e^{\Sigma_1(t-s)}\left(\Sigma_2\xi(s) + \Sigma_3\xi(t_k)\right)\mathrm{d}s \\
&= \int_{t_k}^{t} e^{\Sigma_1(t-s)}\left(\Sigma_2\xi(s) + \Sigma_3\xi(t_k)\right)\mathrm{d}s. \tag{10.8}
\end{aligned}$$

Define $\psi(t_k) = \sqrt{\frac{\lambda_{\max}(P)}{\lambda_{\min}(P)}\|\xi(t_k)\|^2 + \frac{\varepsilon}{\delta\lambda_{\min}(P)}}$. Then, it follows from (10.7) and (10.8) that

$$\|e(t)\| \leq \int_{t_k}^{t} e^{\|\Sigma_1\|(t-s)}\left(\|\Sigma_2\|\|\xi(s)\| + \|\Sigma_3\|\|\xi(t_k)\|\right)\mathrm{d}s.$$

If $\|\Sigma_1\| \neq 0$, we have

$$\|e(t)\| \leq \frac{\|\Sigma_2\|\psi(t_k) + \|\Sigma_3\|\|\xi(t_k)\|}{\|\Sigma_1\|}\left(e^{\|\Sigma_1\|(t-t_k)} - 1\right).$$

Note that the next updating occurs before $\|e(t)\|^2 = \varepsilon + \varrho\|u(t)\|^2$ according to the definition of UIS (10.4). Therefore, the lower bound on the inter-event interval $t_{\min}^l$ can be determined by

$$\frac{\|\Sigma_2\|\psi(t_k) + \|\Sigma_3\|\|\xi(t_k)\|}{\|\Sigma_1\|}\left(e^{\|\Sigma_1\|t_{\min}^l} - 1\right) = \sqrt{\varepsilon + \varrho\xi^T(t)\bar{C}^T\bar{C}\xi(t)} \geq \sqrt{\varepsilon},$$

which means that $e^{\|\Sigma_1\|t_{\min}^l} \geq 1 + \frac{\sqrt{\varepsilon}}{\Delta(t_k)}$, where $\Delta(t_k) = \frac{\|\Sigma_2\|\psi(t_k)+\|\Sigma_3\|\|\xi(t_k)\|}{\|\Sigma_1\|}$. Note that conditions $\|\Sigma_1\| > 0$ and $\frac{\sqrt{\varepsilon}}{\Delta(t_k)} > 0$ hold, which indicate that for any given updating instant t_k, $t_{\min}^l > 0$ always holds.

If $\|\Sigma_1\| = 0$, we have $\|e(t)\| \leq (t - t_k)(\|\Sigma_2\|\psi(t_k) + \|\Sigma_3\|\|\xi(t_k)\|)$. As aforementioned before, since the next updating will not occur before $\|e(t)\|^2 = \varepsilon + \varrho\|u(t)\|^2$, the lower bound on the inter-event interval $t_{\min}^l$ can be determined by $t_{\min}^l(\|\Sigma_2\|\psi(t_k) + \|\Sigma_3\|\|\xi(t_k)\|) = \sqrt{\varepsilon + \varrho\xi^T(t)\bar{C}^T\bar{C}\xi(t)}$, which means that $t_{\min}^l(\|\Sigma_2\|\psi(t_k) + \|\Sigma_3\|\|\xi(t_k)\|) \geq \sqrt{\varepsilon}$, which also implies that $t_{\min}^l > 0$.

From the above discussions, a positive lower bound of the minimum inter-event interval always exists, which completes the proof. □

Combining with the stability condition proposed in Theorem 10.1, a solution of the controller parameters K_1, K_2, K_3, and K_4 is presented in the following theorem.

Theorem 10.3. *For system (10.1) with the event-triggered inputs, there always exists a DOF controller in the form of (10.2) such that states of the closed-loop control system (10.3) are globally UUB, if there exist matrices W, R, L, F, $X > 0$, and $Y > 0$, satisfying*

$$\begin{bmatrix} \Xi_{11} + \Xi_{11}^T & W + \Xi_{21}^T & -XB & \sqrt{\varrho}C^TR^T \\ * & \Xi_{22} + \Xi_{22}^T & -B & \sqrt{\varrho}F^T \\ * & * & -I & 0 \\ * & * & * & -I \end{bmatrix} < 0 \qquad (10.9)$$

and $I - YX < 0$, where $\Xi_{11} = XA + LC$, $\Xi_{21} = A + BRC$, $\Xi_{22} = AY + BF$.

Furthermore, if the above conditions are satisfied, parameters of the desired DOF controller can be chosen as

$$K_1 = U^{-T}(W - XAY - XBF - LCY + XBRCY)V^{-T},$$
$$K_2 = U^{-T}L - U^{-T}XBR, \quad K_3 = FV^{-T} - RCYV^{-T}, \quad K_4 = R, \qquad (10.10)$$

where V and U are two any nonsingular matrices satisfying $I - YX = VU$.

Proof. Define matrix $P = T_2T_1^{-1}$, where T_1 and T_2 are given as

$$T_1 = \begin{bmatrix} I & Y \\ 0 & V^T \end{bmatrix}, \quad T_2 = \begin{bmatrix} X & I \\ U & 0 \end{bmatrix},$$

where $YX + VU = I$, and U, V are any nonsingular matrices. It can be verified that

$$T_1^{-1} = \begin{bmatrix} I & -YV^{-T} \\ 0 & V^{-T} \end{bmatrix}, \quad P = \begin{bmatrix} X & U^T \\ U & -UYV^{-T} \end{bmatrix}.$$

Note that $YX - I > 0$ holds, thus $VUY = Y(Y^{-1} - X)Y < 0$ and $-UYV^{-T} = -V^{-1}Y(Y^{-1} - X)YV^{-T} > 0$. Moreover, referring to the well-known Schur complement lemma, it is obtained that $X - U^T(-UYV^{-T})^{-1}U = X + U^T V^T Y^{-1} = Y^{-1} > 0$, which implies that $P > 0$.

With the DOF controller parameters K_i, $i = 1, 2, 3, 4$, it can be shown that

$$T_2^T \bar{A} T_1 = \begin{bmatrix} \Xi_{11} & W \\ \Xi_{21} & \Xi_{22} \end{bmatrix}, \quad T_2^T \bar{B} = \begin{bmatrix} -XB \\ -B \end{bmatrix},$$

$$T_1^T \bar{C}^T \bar{C} T_1 = \begin{bmatrix} C^T R^T RC & C^T R^T F \\ * & F^T F \end{bmatrix}.$$

Thus, by the Schur complement lemma, the inequality (10.9) is rewritten as

$$\begin{bmatrix} T_1^T \bar{A}^T T_2 + T_2^T \bar{A} T_1 + \varrho T_1^T \bar{C}^T \bar{C} T_1 & T_2^T \bar{B} \\ \bar{B}^T T_2 & -I \end{bmatrix} < 0,$$

and notice that

$$\begin{bmatrix} \bar{A}^T P + P\bar{A} + \varrho \bar{C}^T \bar{C} & P\bar{B} \\ \bar{B}^T P & -I \end{bmatrix}$$

$$= \begin{bmatrix} T_1^{-T} & 0 \\ 0 & I \end{bmatrix} \begin{bmatrix} T_1^T \bar{A}^T T_2 + T_2^T \bar{A} T_1 + \varrho T_1^T \bar{C}^T \bar{C} T_1 & T_2^T \bar{B} \\ \bar{B}^T T_2 & -I \end{bmatrix} \begin{bmatrix} T_1^{-1} & 0 \\ 0 & I \end{bmatrix}.$$

we can conclude that the inequality (10.5) holds, and the states of the closed-loop system (10.3) are globally UUB. $\qquad\square$

10.4 EVENT-TRIGGERED DOF CONTROL FOR SYSTEMS WITH SAMPLED OUTPUT MEASUREMENTS

In this section, we consider that the system output is sampled periodically with a constant sampling period $T_s > 0$. In other words, only the sampled output measurements $y(iT_s)$, $i \in \mathbb{N}$, are available for controller (10.2). In such a case, the DOF controller (10.2) is modified as

$$\dot{\hat{x}}_c(t) = K_1 \hat{x}_c(t) + K_2 \hat{x}_c(iT_s) + K_3 y(iT_s),$$
$$u(t) = K_4 \hat{x}_c(t) + K_5 \hat{x}_c(iT_s) + K_6 y(iT_s), \tag{10.11}$$

where $\hat{x}_c(t) \in \mathbb{R}^n$ denotes the controller state, $\hat{x}_c(iT_s)$ denotes the controller state at sampled instants and $y(iT_s)$ denotes the sampled output measurement. In this situation, the controller (10.11) will receive the sampled output signal $y(iT_s)$ at the sampling instant iT_s, and the $y(iT_s)$ will be held until next sampling instant $(i+1)T_s$.

In Section 10.3, the control input of the plant will be updated at t_k defined in (10.4). To determine the updating instants t_k in (10.4), the continuous supervision of $e(t)$ is required. As shown in Figure 10.1, the UIS is assumed to work in a discrete manner here, and monitor the control signal $u(t)$ with a constant sampling period T_s. In this situation, we can determine the updating instants by $t_k = i_k T_s$, where i_k

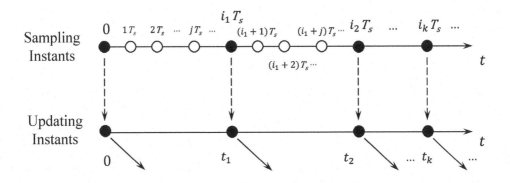

Figure 10.1   Illustration of discrete UIS

are some integers and $\{i_0, i_1, i_2, \ldots\} \subset \{0, 1, 2, 3, \ldots\}$ with $i_0 = 0$ and $i_k < i_{k+1}$. To hold the control input signal continuous, a ZOH is embedded, thus $u(t_k)$ will be held during the time-interval $[t_k, t_{k+1})$. Define $l_{k,j} = (i_k + j)T_s$, $j = 0, 1, 2, \ldots, i_{k+1} - i_k - 1$, which leads to $[t_k, t_{k+1}) = \cup_{j=0}^{i_{k+1}-i_k-1}[l_{k,j}, l_{k,j+1})$. Let $\tau(t) = t - l_{k,j}$, and $\tau(t)$ satisfies $0 \le \tau(t) \le T_s$ with $t \in [l_{k,j}, l_{k,j+1})$, obviously. Furthermore, we define $e_\sigma(t) = u(l_{k,j}) - u(i_k T_s)$, $j = 1, 2, \ldots$, thus $e_\sigma(t)$ is a continuous from the right and piecewise constant function. Above all, the updating instant $i_{k+1} T_s$ can be determined by

$$i_{k+1} = \max_{m \in \mathbb{Z}}\{m > i_k | \|e_\sigma(mT_s)\|^2 \le \bar{\varepsilon} + \bar{\varrho}\|u(mT_s)\|^2\}, \tag{10.12}$$

where $\bar{\varepsilon}$ and $\bar{\varrho}$ are the event-triggering condition parameters chosen in advance. It can be verified that, inequality $\|e_\sigma(t)\|^2 \le \bar{\varepsilon} + \bar{\varrho}\|u(t)\|^2$ always holds for $t \in [i_k T_s, i_{k+1} T_s)$.

For time $t \in [l_{k,j}, l_{k,j+1})$, the plant and controller system can be reorganized as:

$$\dot{x}(t) = Ax(t) + Bu(t_k)$$
$$= Ax(t) + Bu(l_{k,j}) - Bu(l_{k,j}) + Bu(t_k)$$
$$= Ax(t) + B(K_4 + K_5)\hat{x}_c(l_{k,j}) + BK_6 y(l_{k,j}) - Be_\sigma(t)$$
$$= Ax(t) + B(K_4 + K_5)\hat{x}_c(t - \tau(t)) + BK_6 Cx(t - \tau(t)) - Be_\sigma(t),$$
$$\dot{\hat{x}}_c(t) = K_1\hat{x}_c(t) + K_2\hat{x}_c(l_{k,j}) + K_3 y(l_{k,j})$$
$$= K_1\hat{x}_c(t) + K_2\hat{x}_c(t - \tau(t)) + K_3 Cx(t - \tau(t)).$$

Define variable $\xi(t) = [x^T(t), \hat{x}_c^T(t)]^T$ and the closed-loop system can be determined as

$$\dot{\xi}(t) = \tilde{A}\xi(t) + \tilde{A}_d\xi(t - \tau(t)) + \tilde{B}e_\sigma(t), \ t \in [l_{k,j}, l_{k,j+1}), \tag{10.13}$$

where $\tilde{A} = \begin{bmatrix} A & 0 \\ 0 & K_1 \end{bmatrix}$, $\tilde{B} = \begin{bmatrix} -B \\ 0 \end{bmatrix}$, $\tilde{A}_d = \begin{bmatrix} BK_6 C & B(K_4 + K_5) \\ K_3 C & K_2 \end{bmatrix}$.

Before studying the stability of closed-loop system (10.13), the following lemma is given in advance.

Lemma 10.1. *[104] Let $x(t)$: (p,q) $\rightarrow\in$ $\mathbb{R}^n$ be absolutely continuous with $\dot{x}(t)$ $\in$ $\mathcal{L}_2(p,q)$ and $x(p)$ $=$ 0, and then inequality $\int_p^q x^T(s)Mx(s)\mathrm{d}s$ $\leq$ $\frac{4(q-p)^2}{\pi^2}\int_p^q \dot{x}^T(s)M\dot{x}(s)\mathrm{d}s$ holds for any $M > 0$.*

Motivated by [104], we can arrive at the following results.

Theorem 10.4. *Consider the closed-loop system (10.13) with updating instants determined by*

$$t_{k+1} = i_{k+1}T_s, \tag{10.14}$$

where i_k is given by (10.12) and T_s denotes the sampling period. For a given parameter $\bar{\varrho} > 0$, if there exist matrices N_1, N_2, $P > 0$, and $S > 0$ with appropriate dimensions satisfying the following matrix inequality,

$$\tilde{\Phi} \triangleq \Phi + N^T \tilde{B}\tilde{B}^T N + \bar{\varrho}\tilde{C}^T\tilde{C} < 0, \tag{10.15}$$

where

$$\Phi = \begin{bmatrix} \Phi_1 & P - N_1^T + (\tilde{A} + \tilde{A}_d)^T N_2 & -N_1^T\tilde{A}_d \\ * & T_s^2 S - N_2 - N_2^T & -N_2^T\tilde{A}_d \\ * & * & -\frac{\pi^2}{4}S \end{bmatrix},$$

$$\tilde{C} = \begin{bmatrix} K_6C & K_4 + K_5 & 0 & 0 & -K_6C & -K_5 \end{bmatrix}, \quad N = \begin{bmatrix} N_1 & N_2 & 0 \end{bmatrix},$$

with $\Phi_1 = N_1^T(\tilde{A} + \tilde{A}_d) + (\tilde{A} + \tilde{A}_d)^T N_1$, then the states of system (10.13) are globally UUB.

Proof. Consider the following Lyapunov functional for $t \in [l_{k,j}, l_{k,j+1})$,

$$V(t) = \xi^T(t)P\xi(t) + T_s^2 \int_{t-\tau(t)}^t \dot{\xi}^T(s)S\dot{\xi}(s)\mathrm{d}s - \frac{\pi^2}{4}\int_{t-\tau(t)}^t \nu^T(s)S\nu(s)\mathrm{d}s,$$

where $P > 0$, $S > 0$, and $\nu(t) = \xi(t) - \xi(t - \tau(t))$. It turns out that $\dot{\tau}(t) = 1, \dot{\xi}(t - \tau(t)) = 0$, thus $\dot{\nu}(t) = \dot{\xi}(t)$, and it follows from the Wirtinger inequality, $V(t) \geq 0$ and the last two terms in $V(t)$ vanishes at $t = l_{k,j}$, i.e., $V(l_{k,j}) = \xi^T(l_{k,j})P\xi(l_{k,j})$. Hence, condition $\lim_{t\to l_{k,j}} V(t) \geq V(l_{k,j})$ holds.

Differentiating $V(t)$ along system (10.13), we have

$$\dot{V}(t) = 2\xi^T(t)P\dot{\xi}(t) + T_s^2\dot{\xi}^T(t)S\dot{\xi}(t) - \frac{\pi^2}{4}\nu^T(t)S\nu(t),$$

which together with the fact

$$2(\xi^T(t)N_1^T + \dot{\xi}^T(t)N_2^T)((\tilde{A} + \tilde{A}_d)\xi(t) - \tilde{A}_d\nu(t) + \tilde{B}e_\sigma(t) - \dot{\xi}(t)) = 0$$

gives that $\dot{V}(t) \leq \eta^T(t)\Phi\eta(t) + 2(\xi^T(t)N_1^T + \dot{\xi}^T(t)N_2^T)\tilde{B}e_\sigma(t)$, where $\eta(t) = [\xi^T(t), \dot{\xi}^T(t), \nu^T(t)]^T$. Note that

$$2(\xi^T(t)N_1^T + \dot{\xi}^T(t)N_2^T)\tilde{B}e_\sigma(t) \leq \eta^T(t)\begin{bmatrix} N_1^T \\ N_2^T \\ 0 \end{bmatrix}\tilde{B}\tilde{B}^T\begin{bmatrix} N_1 & N_2 & 0 \end{bmatrix}\eta(t) + \|e_\sigma(t)\|^2,$$

which together with $\|e_\sigma(t)\|^2 \leq \bar{\varepsilon} + \bar{\varrho}\eta^T(t)\tilde{C}^T\tilde{C}\eta(t)$ yields $\dot{V}(t) \leq \eta^T(t)\tilde{\Phi}\eta(t) + \bar{\varepsilon}$. Thus, the states of the closed-loop system (10.13) are shown to be globally UUB, which completes the proof. □

With the stability condition presented in Theorem 10.4, the following theorem presents a solution of the controller parameters K_i, $i = 1, 2, 3, 4, 5, 6$.

Theorem 10.5. *For system (10.1) with event-triggered inputs, there exists a DOF controller in the form of (10.11) such that the states of closed-loop control system (10.13) are globally UUB, if there exist matrices W, R, L, F, H, Z, $X > 0$ and $Y > 0$ satisfying*

$$\begin{bmatrix} \Lambda_{11} & \Lambda_{12} & \Lambda_{13} \\ * & \Lambda_{22} & \Lambda_{23} \\ * & * & \Lambda_{33} \end{bmatrix} < 0 \tag{10.16}$$

and $I - YX < 0$, where

$$\Lambda_{11} = \begin{bmatrix} \Xi_{11} + \Xi_{11}^T & H + \Xi_{21}^T \\ * & \Xi_{22} + \Xi_{22}^T \end{bmatrix}, \quad \Lambda_{12} = \begin{bmatrix} \Xi_{11}^T & \Xi_{21}^T & -LC & -W \\ H^T & \Xi_{22}^T & -BRC & -BF \end{bmatrix},$$

$$\Lambda_{13} = \begin{bmatrix} -XB & \sqrt{\bar{\varrho}}C^T R^T \\ -B & \sqrt{\bar{\varrho}}F^T \end{bmatrix}, \quad \Lambda_{22} = \begin{bmatrix} (T_s^2 - 2)X & (T_s^2 - 2)I & -LC & -W \\ * & (T_s^2 - 2)Y & -BRC & -BF \\ * & * & -\frac{\pi^2}{4}X & -\frac{\pi^2}{4}I \\ * & * & * & -\frac{\pi^2}{4}Y \end{bmatrix},$$

$$\Lambda_{23} = \begin{bmatrix} -XB & 0 \\ -B & 0 \\ 0 & \sqrt{\bar{\varrho}}C^T R^T \\ 0 & \sqrt{\bar{\varrho}}Z^T \end{bmatrix}, \quad \Lambda_{33} = \begin{bmatrix} -I & 0 \\ * & -I \end{bmatrix},$$

with $\Xi_{11} = XA + LC$, $\Xi_{21} = A + BRC$, and $\Xi_{22} = AY + BF$.

Furthermore, if the above conditions are satisfied, parameters of the desired DOF controller can be chosen as

$$K_1 = U^{-T}(H - W - XAY)V^{-T},$$

$$K_2 = U^{-T}(W - XBF - LCY + XBRCY)V^{-T}, K_3 = U^{-T}(L - XBR),$$

$$K_4 = (F - Z)V^{-T}, K_5 = (Z - RCY)V^{-T}, \ K_6 = R, \tag{10.17}$$

where V and U are any nonsingular matrices satisfying $VU = YX + I$.

Proof. With the same definitions for T_1 and T_2 in the proof of Theorem 10.3, letting $P = S = N_1 = N_2 = T_2T_1^{-1}$, we can arrive at $P > 0$ and $S > 0$. Moreover, with the DOF controller parameters K_i, $i = 1, 2, \ldots, 6$, it can be obtained that

$$T_2^T(\tilde{A} + \tilde{A}_d)T_1 = \begin{bmatrix} XA + LC & H \\ A + BRC & AY + BF \end{bmatrix}, \ T_2^T\tilde{A}_dT_1 = \begin{bmatrix} LC & W \\ BRC & BF \end{bmatrix},$$

$$T_1^T T_2 = T_2^T T_1 = \begin{bmatrix} X & I \\ I & Y \end{bmatrix}, \ T_2^T\tilde{B} = \begin{bmatrix} -XB \\ -B \end{bmatrix},$$

$$\tilde{C}_1 T_1 = \begin{bmatrix} RC & F \end{bmatrix}, \ \tilde{C}_3 T_1 = \begin{bmatrix} -RC & -Z \end{bmatrix},$$

where $\tilde{C}_1 = [K_6C \ \ K_4 + K_5]$ and $\tilde{C}_3 = [-K_6C \ \ - K_5]$. By the Schur complement lemma, inequality (10.15) can be rewritten as

$$
\tilde{\Phi} = \begin{bmatrix}
\mathcal{A}_1 & P - N_1^T + (\tilde{A} + \tilde{A}_d)^T N_2 & -N_1^T \tilde{A}_d & N_1^T \tilde{B} & \sqrt{\varrho}\tilde{C}_1^T \\
* & T_s^2 S - N_2 - N_2^T & -N_2^T \tilde{A}_d & N_2^T \tilde{B} & 0 \\
* & * & -\frac{\pi^2}{4}S & 0 & \sqrt{\varrho}\tilde{C}_3^T \\
* & * & * & -I & 0 \\
* & * & * & * & -I
\end{bmatrix} < 0
$$

and $\mathcal{A}_1 = N_1^T(\tilde{A} + \tilde{A}_d) + (\tilde{A} + \tilde{A}_d)^T N_1$. Then, we have

$$
\tilde{\Phi} = diag\{T_1^{-T}, T_1^{-T}, T_1^{-T}, I, I\}\bar{\Phi}diag\{T_1^{-1}, T_1^{-1}, T_1^{-1}, I, I\},
$$

where

$$
\bar{\Phi} = \begin{bmatrix}
\mathcal{A}_2 & T_1^T(\tilde{A} + \tilde{A}_d)^T T_2 & -T_2^T \tilde{A}_d T_1 & T_2^T \tilde{B} & \sqrt{\varrho}T_1^T \tilde{C}_1^T \\
* & (T_s^2 - 2)T_1^T T_2 & -T_2^T \tilde{A}_d T_1 & T_2^T \tilde{B} & 0 \\
* & * & -\frac{\pi^2}{4}T_1^T T_2 & 0 & \sqrt{\varrho}T_1^T \tilde{C}_3^T \\
* & * & * & -I & 0 \\
* & * & * & * & -I
\end{bmatrix} < 0.
$$

and $\mathcal{A}_2 = T_2^T(\tilde{A} + \tilde{A}_d)T_1 + T_1^T(\tilde{A} + \tilde{A}_d)^T T_2$. Thus, inequality (10.16) is obtained, and the states of closed-loop control system (10.13) are globally UUB, which completes the proof. □

10.5 NUMERICAL SIMULATION

In this section, the proposed control approaches will be applied to the satellite system considered in Chapter 5. The initial state $x(0)$ is assumed to be $[0.2 \ 0.3 \ -0.3 \ -0.2]^T$, and then event-triggering condition parameters are given as $\sqrt{\varepsilon} = 0.005$ and $\sqrt{\varrho} = 0.005$. From Theorem 10.3, the controller parameter matrices K_1, K_2, K_3, and K_4 can be computed as

$$
K_1 = \begin{bmatrix}
50.3502 & 2.8629 & 28.1404 & 5.9067 \\
-50.6963 & -2.967 & -31.179 & -5.0432 \\
-80.7737 & -4.5354 & -44.266 & -9.5086 \\
-107.8983 & -7.2436 & -58.7651 & -13.2154
\end{bmatrix},
$$

$$
K_2 = \begin{bmatrix} 1.2551 & -17.8524 & 1.2157 & -37.7471 \end{bmatrix}^T,
$$

$$
K_3 = \begin{bmatrix} -80.6856 & 4.5995 & 43.4514 & 9.5605 \end{bmatrix}, \quad K_4 = -0.1168.
$$

The simulation results are shown in Figure 10.2. Figure 10.2(a–c) shows the states and the control input of closed-loop system with the traditional continuous-time control scheme and proposed event-triggered control scheme. It can be seen that the performance of system with the event-triggered control scheme is almost same as the traditional continuous-time control one, and states of closed-loop system are globally UUB. Furthermore, both the error signal $\|e(t)\|$ and inter-event intervals are shown in

Figure 10.2(d). It turns out from Figure 10.2(d) that the communication frequency can be highly reduced for the closed-loop system with the event-triggered DOF controller.

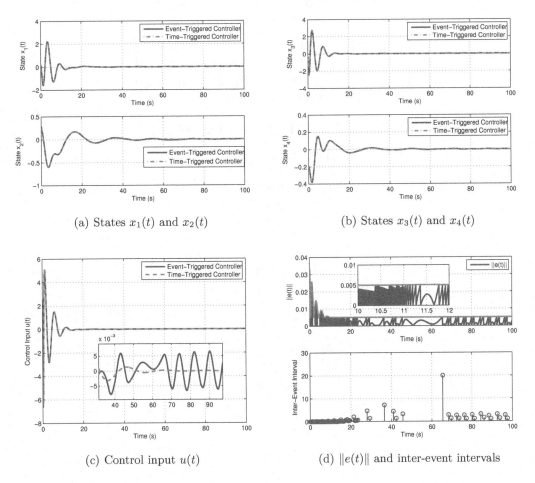

(a) States $x_1(t)$ and $x_2(t)$

(b) States $x_3(t)$ and $x_4(t)$

(c) Control input $u(t)$

(d) $\|e(t)\|$ and inter-event intervals

Figure 10.2 Comparative simulation results of time- and event-triggered DOF control approaches (the continuous output case)

Next, the simulation results of the proposed event-triggered control approach will be shown for system with sampled outputs.

Take the sampling period $T_s = 0.015$s. The initial state $x(0)$ is also assumed as $[0.2 \ 0.3 \ -0.3 \ -0.2]^T$, and then event-triggering condition parameters are given by $\sqrt{\bar{\varepsilon}} = 0.001$ and $\sqrt{\bar{\varrho}} = 0.01$. From Theorem 10.5, the controller parameter matrices

K_i, $i = 1, 2, \ldots, 6$ can be computed as

$$K_1 = \begin{bmatrix} -0.3721 & 0.0769 & 0.9076 & -0.0577 \\ 9.5757 & 2.2631 & 1.577 & 2.0642 \\ -0.4246 & 0.0074 & -1.0244 & 0.006 \\ -2.2813 & -2.0573 & -4.5547 & -0.9468 \end{bmatrix},$$

$$K_2 = \begin{bmatrix} 6.4422 & -0.022 & 3.2386 & 0.6606 \\ -6.072 & -2.3385 & 0.7294 & -1.5865 \\ -8.4414 & -0.0294 & -4.1502 & -0.8894 \\ -27.8418 & 1.0270 & -15.4902 & -2.4742 \end{bmatrix},$$

$$K_3 = \begin{bmatrix} -1.9098 & -5.1231 & 2.3316 & 10.9882 \end{bmatrix}^T,$$

$$K_4 = \begin{bmatrix} 3.5304 & 0.3441 & 1.2059 & 0.5074 \end{bmatrix},$$

$$K_5 = \begin{bmatrix} 5.105 & -0.3181 & 3.0463 & 0.4008 \end{bmatrix}, \quad K_6 = -2.3973.$$

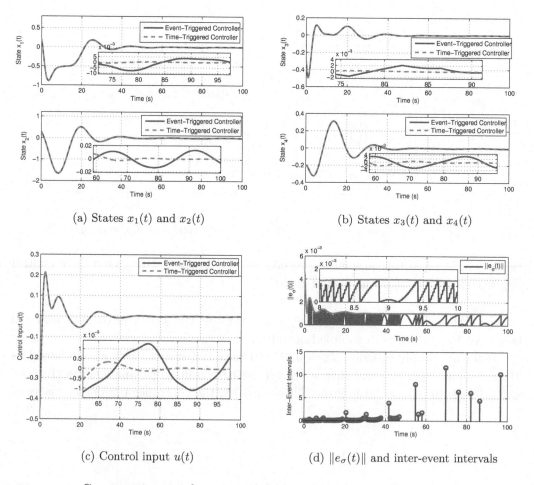

(a) States $x_1(t)$ and $x_2(t)$

(b) States $x_3(t)$ and $x_4(t)$

(c) Control input $u(t)$

(d) $\|e_\sigma(t)\|$ and inter-event intervals

Figure 10.3 Comparative simulation results of time- and event-triggered DOF control approaches (the sampled output case)

Then, the simulation results are shown in Figure 10.3. Obviously, with the designed dynamic sampled output feedback controller (10.11), the states of closed-loop system are globally UUB, and the effectiveness of the proposed event-triggered control approach is illustrated.

10.6 SUMMARY

This chapter has investigated the problem of DOF control of systems with event-triggered inputs, and the control schemes have been proposed for systems with continuous and sampled output measurements, respectively. The updating instants and the controller parameters have been designed. It has been illustrated that with the designed UIS, the states of the resulting closed-loop system are UUB. Finally, a numerical example have been presented to illustrate the applicability of the developed results.

Co-Design of Event-Triggered Control and Quantized Control for CPSs

11.1 INTRODUCTION

In CPSs, the measurements and control signals are exchanged through a capacity-limited communication network. Due to the limited network capacity or limited transmission data rate, the communication network may have considerable influences on the performances of the closed-loop systems. Generally, there are two possible ways to alleviate such influences on the control performance, one is to develop event-triggered strategies to reduce the communication frequency over the communication network, and the other is to employ signal quantization schemes to convert the real-valued signal into piecewise constant signal with finite values. In view of the appealing advantages of the event-triggered control and quantized control, it is intuitive to consider the combination of both two approaches, and there are some interesting results on this topic recently [106, 47, 127, 136, 222, 170]. In [47], the model-based event-triggered control problems have been considered for systems with quantization and network communication delays. In [89], on the basis of the event-triggered control schemes proposed in [106], the quantized event-triggered control schemes are proposed, where the quantized state information at the updating instants is required to generate the control inputs. In [127], the event-triggered adaptive differential modulation has been proposed, which combines the event-triggered strategy and data coding technique for network capacity reduction. The event-triggered H_∞ control approach is proposed for singular systems with the quantizations of both the states and control inputs [136] by using a time-delay system approach. In [222, 170], the event-triggered consensus control approaches are established for multi-agent systems.

In this chapter, we further concentrate on the combination of event-triggered control and quantized control techniques, and designing the quantized event-triggered controller for both the fixed and relative event threshold cases. First, similar to the predictive vector quantizer (PVQ) [50], both the encoder and decoder, the UIS and event-triggered quantizer are designed for the event-triggered control system. It is worth mentioning that the proposed quantizer only requires a few parameters, rather than the plant's specific model. Then, the stability analysis problems are addressed for the quantized event-triggered control systems with both fixed and relative event thresholds, respectively. For the fixed event threshold case, the range of quantization length is obtained and then the states of closed-loop system are globally UUB. For the relative event threshold case, the modified quantizer and UIS are proposed, and the global asymptotic stability of the closed-loop system is guaranteed. Furthermore, the existence on the minimum inter-event interval to exclude the Zeno behavior is shown for both the two cases. Finally, the validity of the proposed approaches is illustrated by the numerical simulations.

11.2 SYSTEM DESCRIPTION

The quantized event-triggered control system is shown in Figure 11.1, the controlled plant considered in this chapter is described by

$$\dot{x}(t) = Ax(t) + Bu(t) \tag{11.1}$$

where $x(t) \in \mathbb{R}^n$ and $u(t) \in \mathbb{R}^m$ are the state and input vectors, respectively. A, B, and C are constant system matrices with appropriate dimensions, and the initial state is given by $x(t_0) = x_0$.

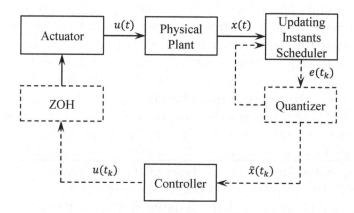

Figure 11.1 Block diagram of quantized event-triggered control systems

The controller here is described as

$$u(t) = -K\tilde{x}(t_k), \ t \in [t_k, t_{k+1}), \ k \in \mathbb{N} \tag{11.2}$$

where $K \in \mathbb{R}^{m \times n}$ is the state feedback controller gain, $\tilde{x}(t_k) \in \mathbb{R}^n$ is output of the quantizer, which denotes the quantized value of state $x(t_k)$, where t_k is the updating instant generated by the following UIS,

$$t_{k+1} = \sup\{t > t_k |\ \zeta(t) \le \alpha\}, \ k \in \mathbb{N} \tag{11.3}$$

where α is a constant event threshold chosen by the designer and $\zeta(t)$ is the event-triggering function to be determined later. It is easy to notice that as the controller is event-triggered, and the control input is only updated at the updating instant t_k, and $\bigcup_{k \in \mathbb{N}}[t_k, t_{k+1}) = [t_0, \infty)$.

As shown in Figure 11.1, the UIS is designed to monitor state $x(t)$ continuously and determine the updating instants $\{t_k\}_{k=0}^{\infty}$. At each updating instant, the error signal $e(t)$ defined by

$$e(t) = x(t) - \tilde{x}(t_k), \ t \in [t_k, t_{k+1}), \ k \in \mathbb{N} \tag{11.4}$$

will be sent to the encoder. The controller determines the control input $u(t_k)$ based on quantized state $\tilde{x}(t_k)$, and the ZOH is implemented to keep the control signal to be continuous.

In this chapter, the quantized event-triggered control problem for continuous-time systems is considered. More specifically, we will design the event-triggered quantizer for both fixed and relative event thresholds, respectively, including the encoder and the decoder, and also select the suitable event-triggering function $\zeta(t)$ based on the quantized states. Moreover, the performances of the closed-loop system are analyzed, including the stability analysis and the existence of the lower bound of the minimum inter-event interval.

11.3 QUANTIZED EVENT-TRIGGERED CONTROL: FIXED EVENT THRESH-OLD STRATEGY

In this section, we focus on the quantized event-triggered control with fixed event threshold. Motivated by [50], the event-triggered encoder and decoder are proposed firstly, and then the UIS and the quantizer are also designed.

For the fixed event threshold case, the event-triggering function $\zeta(t)$ can be simply defined based on the quantized state $\tilde{x}(t_k)$ as

$$\zeta(t) = \|e(t)\| = \|x(t) - \tilde{x}(t_k)\|, \ t \in [t_k, t_{k+1}), \ k \in \mathbb{N} \tag{11.5}$$

where $\tilde{x}(t_k) \in \mathbb{R}^n$ is the quantized state at instant t_k as mentioned before. It is easy to see that the event-triggering function $\zeta(t)$ is a piecewise continuous function. According to (11.3) and (11.5), the inequality $\|x(t) - \tilde{x}(t_k)\| < \alpha$ holds for $t \in [t_k, t_{k+1})$, so $\|x(t_k) - \tilde{x}(t_k)\| < \alpha$, and at t_{k+1}, we have $\|x(t_{k+1}) - \tilde{x}(t_k)\| = \alpha$.

Remark 11.1. *In the existing literatures, such as [198] the event-triggering functions are always defined as $\zeta(t) = \|x(t) - x(t_k)\|$ and $\zeta(t_k) \equiv 0$ always holds. However, in this chapter, the event-triggering function $\zeta(t)$ is defined based on the quantized state $\tilde{x}(t_k)$ instead of $x(t_k)$. In this situation, the current state $x(t)$ is compared with the quantized state $\tilde{x}(t_k)$ constantly, and the quantized state $\tilde{x}(t_k)$ will be replaced by new quantized state $\tilde{x}(t_{k+1})$ at instant t_{k+1}. Hence, the event-triggering function $\zeta(t)$ cannot be reset to zero at the each updating instants, i.e., $\zeta(t_k) \not\equiv 0$ [47].*

The event-triggered quantizer consists of event-triggered encoder and decoder, and at the initial time t_0, both the event-triggered encoder and decoder store the quantized state $\tilde{x}(t_0)$. The structure of the event-triggered encoder is shown in Figure 11.2, where the encoder processes the error signal $e(t_{k+1}^-)$, and sends the index number q_{k+1} to the decoder. Here, q_{k+1} is determined by

$$q_{k+1} = \min \left\{ i \;\middle|\; \min_{i \in \aleph} \left\| e(t_{k+1}^-) - \Delta M(i)\mathbf{1} \right\| \right\}, \tag{11.6}$$

where $\mathbf{1} = [1\ 0\ \cdots\ 0]^T$, $\aleph = \{0, 1, 2, \ldots, 2n-1\}$, Δ denotes the quantization length, and $M(i)$ is given by

$$M(i) = \begin{bmatrix} 0 & (-1)^n \\ -I_{n-1} & 0 \end{bmatrix}^i. \tag{11.7}$$

According to (11.6), the quantized value of $e(t_{k+1}^-)$ can be defined as

$$\tilde{e}(t_{k+1}^-) = \Delta M(q_{k+1})\mathbf{1}, \tag{11.8}$$

which together with $\tilde{x}(t_k)$ determines $\tilde{x}(t_{k+1})$ according to

$$\tilde{x}(t_{k+1}) = \tilde{x}(t_k) + \tilde{e}(t_{k+1}^-), \quad k \in \mathbb{N}. \tag{11.9}$$

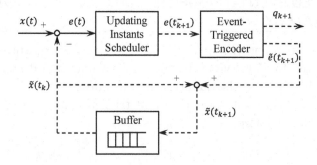

Figure 11.2 Structure of the event-triggered encoder

Figure 11.3 demonstrates the structure of the event-triggered decoder. Once the decoder receives the sequence of index number q_{k+1}, the decoder converts it into the quantized error signal $\tilde{e}(t_{k+1}^-)$ according to (11.8). Since $\tilde{x}(t_k)$ has already been stored in the buffer, at the decoder side, the quantized states $\tilde{x}(t_{k+1})$ can be determined by $\tilde{x}(t_{k+1}) = \Delta M(q_{k+1})\mathbf{1} + \tilde{x}(t_k)$, $k \in \mathbb{N}$.

Remark 11.2. *It should be noticed that, for the proposed event-triggered quantizer, the initial quantized state $\tilde{x}(t_0)$ is assumed to be known for both the encoder and decoder. Here, we choose $\tilde{x}(t_0)$ as $x(t_0)$ for simplicity, and $x(t_0)$ will be stored by the encoder and the decoder at the initial time, respectively. Then, the sequence of index numbers $\{q_k\}$ is determined according to (11.6), and the quantized states $\{\tilde{x}(t_k)\}$ can be generated constantly, for $k \geq 1$.*

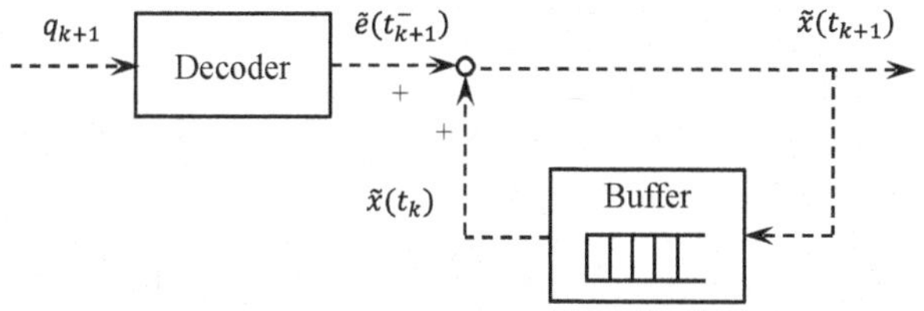

Figure 11.3 Structure of the decoder

To obtain the main results, the following two lemmas are presented.

Lemma 11.1. *There is only one element in* $\Delta M(i)\mathbf{1}$ *is not equal to 0 and* $\|\Delta M(i)\mathbf{1}\| = \Delta$.

Proof. By calculating (11.7), we get

$$
M(i) = \begin{bmatrix} 0 & \rho_1 I_{\psi(i)-1} \\ \rho_2 I_{n+1-\psi(i)} & 0 \end{bmatrix},
$$

where $\rho_1, \rho_2 \in \{-1, 1\}$ and

$$
\psi(i) = \begin{cases} i+1 & \text{if } i \in \{0, 1, \cdots, n-1\}, \\ i+1-n & \text{if } i \in \{n, n+1, \cdots, 2n-1\}. \end{cases}
$$

Thus, we have $\Delta M(i)\mathbf{1} = [0\ 0\ \cdots\ \rho_2\Delta \cdots\ 0]^T$, where $\rho_2\Delta$ is the $\psi(i)$-th element of $\Delta M(i)\mathbf{1}$, which is the only one element that does not equal to zero. Moreover, we can also conclude that $\|\Delta M(i)\mathbf{1}\| = \Delta$. □

Lemma 11.2. *The inequality* $|e_{\psi(q_{k+1})}(t_{k+1}^-)| \geq \frac{\alpha}{\sqrt{n}}$ *holds, where* $e_{\psi(q_{k+1})}(t_{k+1}^-)$ *is the* $\psi(q_{k+1})$-*th element of* $e(t_{k+1}^-)$.

Proof. On the basis of Lemma 11.1, and according to (11.6), we have

$$
\min_{i \in \aleph} \|e(t_{k+1}^-) - \Delta M(i)\mathbf{1}\| = \|e(t_{k+1}^-) - \Delta M(q_{k+1})\mathbf{1}\|
$$

$$
= \left[\alpha^2 + \Delta^2 - 2\Delta |e_{\psi(q_{k+1})}(t_{k+1}^-)| \right]^{\frac{1}{2}}.
$$

On the other hand, we can also obtain

$$
\min_{i \in \aleph} \|e(t_{k+1}^-) - \Delta M(i)\mathbf{1}\| = \left[\alpha^2 + \Delta^2 - 2\Delta \max_{i \in \aleph} |e_{\psi(i)}(t_{k+1}^-)| \right]^{\frac{1}{2}},
$$

which means $|e_{\psi(q_{k+1})}(t_{k+1}^-)| = \max_{i \in \aleph} |e_{\psi(i)}(t_{k+1}^-)| = \max_{j \in \{1,2,\ldots,n\}} |e_j(t_{k+1}^-)|$. Then the inequality $|e_{\psi(q_{k+1})}(t_{k+1}^-)| \geq \frac{\alpha}{\sqrt{n}}$ can be obtained, since $\|e(t_{k+1}^-)\| = \sqrt{\sum_{j=1}^n |e_j(t_{k+1}^-)|^2} = \alpha$. □

The following theorem presents a possible way to choose the quantization length Δ.

Theorem 11.1. *If quantization length Δ is chosen such that $\Delta < \frac{2}{\sqrt{n}}\alpha$, then the error signal $e(t)$ will never exceed the event threshold α.*

Proof. Assume that the error signal $e(t)$ will not exceed the event threshold α, i.e., we always have $\|x(t_{k+1}) - \tilde{x}(t_{k+1})\| < \alpha$, which together with (11.4) and (11.9) gives that

$$
\begin{aligned}
\|x(t_{k+1}) - \tilde{x}(t_{k+1})\| &= \|x(t_{k+1}) - \tilde{x}(t_k) + \tilde{x}(t_k) - \tilde{x}(t_{k+1})\| \\
&= \|e(t_{k+1}^-) - \tilde{e}(t_{k+1}^-)\| \\
&= \|e(t_{k+1}^-) - \Delta M(q_{k+1})\mathbf{1}\| \\
&= \left[\alpha^2 + \Delta^2 - 2\Delta|e_{\psi(q_{k+1})}(t_{k+1}^-)|\right]^{\frac{1}{2}} < \alpha,
\end{aligned}
$$

thus $\Delta < 2|e_{\psi(q_{k+1})}(t_{k+1}^-)|$. According to Lemma 11.2, we have $|e_{\psi(q_{k+1})}(t_{k+1}^-)| \geq \frac{\alpha}{\sqrt{n}}$, therefore, Δ can be selected as $\Delta < \frac{2}{\sqrt{n}}\alpha$. $\qquad\square$

Remark 11.3. *Theorem 11.1 indicates that the quantization length Δ should not be chosen too large, or the error signal $e(t)$ will exceed the event threshold α, which leads to that the UIS is invalid.*

Substituting the controller (11.2) into system (11.1) gives the following closed-loop system,

$$
\begin{aligned}
\dot{x}(t) &= Ax(t) + Bu(t) \\
&= Ax(t) - BK\tilde{x}(t_k) \\
&= (A - BK)x(t) + BK[x(t) - \tilde{x}(t_k)] \\
&= (A - BK)x(t) + BKe(t) \\
&= \bar{A}x(t) + \bar{B}e(t) \qquad\qquad\qquad\qquad (11.10)
\end{aligned}
$$

where $\bar{A} = A - BK$ and $\bar{B} = BK$.

For system (11.10), we have the following theorem.

Theorem 11.2. *Consider the closed-loop system (11.10) and the UIS (11.3) with event-triggering function $\zeta(t)$ given by (11.5). For a given matrix $Q > 0$, if there exists $P > 0$ satisfying the following Riccati equation,*

$$
A^T P + PA - PBB^T P + I_n + Q = 0 \qquad\qquad (11.11)
$$

then the controller gain can be designed as $K = -\frac{1}{2}B^T P$, and the states of system (11.10) are globally UUB and converge into a bounded region $\mathcal{B}(\alpha)$ given by

$$
\mathcal{B}(\alpha) = \left\{ x(t) \big| \|x(t)\|^2 \leq \frac{\alpha^2 \lambda_{\max}(P) \|PBB^T P\|^2}{4\lambda_{\min}(P)\lambda_{\min}(Q)} \right\}.
$$

Proof. Substituting $K = -\frac{1}{2}B^T P$ into the Riccati equation (11.11) yields that

$$\bar{A}^T P + P\bar{A} + I_n + Q = 0.$$

Considering of the Lyapunov function candidate $V(t) = x^T(t)Px(t)$ and the time derivative of $V(t)$, $t \in [t_k, t_{k+1})$, is

$$
\begin{aligned}
\dot{V}(t) &= \dot{x}^T(t)Px(t) + x^T(t)P\dot{x}(t) \\
&= \left(\bar{A}x(t) + \bar{B}e(t)\right)^T Px(t) + x^T(t)P\left(\bar{A}x(t) + \bar{B}e(t)\right) \\
&= -x^T(t)Qx(t) + \|P\bar{B}e(t)\|^2 - \|P\bar{B}e(t) - x(t)\|^2 \\
&\leq -x^T(t)Qx(t) + \|P\bar{B}\|^2\|e(t)\|^2 \\
&\leq -\lambda_{\min}(Q)\|x(t)\|^2 + \|P\bar{B}\|^2\zeta^2(t).
\end{aligned}
\tag{11.12}
$$

According to (11.3) and (11.5), the inequality $\zeta^2(t) \leq \alpha^2$ holds for $t \in [t_k, t_{k+1})$. Thus,

$$
\begin{aligned}
\dot{V}(t) &\leq -\lambda_{\min}(Q)\|x(t)\|^2 + \mu^2 \\
&\leq -\frac{\lambda_{\min}(Q)}{\lambda_{\max}(P)}x^T(t)Px(t) + \mu^2 \\
&= -\sigma V(t) + \mu^2,
\end{aligned}
\tag{11.13}
$$

where $\mu = \alpha\|P\bar{B}\| = \frac{1}{2}\alpha\|PBB^T P\|$ and $\sigma = \frac{\lambda_{\min}(Q)}{\lambda_{\max}(P)}$.

Multiplying $e^{\sigma(t-t_k)}$ from both sides of the inequality (11.13) yields that

$$e^{\sigma(t-t_k)}\dot{V}(t) + e^{\sigma(t-t_k)}\sigma V(t) \leq e^{\sigma(t-t_k)}\mu^2$$

and it can be rewritten as

$$\frac{\mathrm{d}}{\mathrm{d}t}\{e^{\sigma(t-t_k)}V(t)\} \leq e^{\sigma(t-t_k)}\mu^2. \tag{11.14}$$

Calculating the integral of the both sides of (11.14), we get

$$
\begin{aligned}
e^{\sigma(t-t_k)}V(t) - V(t_k) &\leq \mu^2 \int_{t_k}^{t} e^{\sigma(s-t_k)}\,\mathrm{d}s \\
&= \frac{\mu^2}{\sigma}(e^{\sigma(t-t_k)} - 1).
\end{aligned}
$$

Then, the following inequality can be obtained,

$$V(t) \leq e^{-\sigma(t-t_0)}V(t_0) - \frac{\mu^2}{\sigma}e^{-\sigma(t-t_0)} + \frac{\mu^2}{\sigma},$$

which implies that

$$\|x(t)\|^2 \leq \frac{\lambda_{\max}(P)}{\lambda_{\min}(P)}\|x(t_0)\|^2 e^{-\sigma(t-t_0)} + \frac{\mu^2}{\sigma\lambda_{\min}(P)}[1 - e^{-\sigma(t-t_0)}],$$

thus we have

$$\lim_{t \to +\infty} \|x(t)\|^2 \leq \frac{\mu^2}{\sigma \lambda_{\min}(P)} = \frac{\alpha^2 \lambda_{\max}(P) \|PBB^T P\|^2}{4\lambda_{\min}(P)\lambda_{\min}(Q)},$$

which implies that the states of the resulting closed-loop system (11.10) are globally UUB, and will converge to the bounded region $\mathcal{B}(\alpha)$. ☐

In the following theorem, it can be shown that there exists a positive lower bound of the minimum inter-event interval, which implies that the Zeno behavior can be excluded.

Theorem 11.3. *For the updating instants defined by (11.3) with the event-triggering function $\zeta(t)$ in (11.5), there exists a strictly positive lower bound $t_{\min}^l > 0$ such that $t_{\min} \geq t_{\min}^l > 0$.*

Proof. During the time interval $t \in [t_k, t_{k+1})$, we have

$$\begin{aligned} \dot{e}(t) &= Ax(t) + Bu(t) \\ &= Ax(t) - BK\tilde{x}(t_k) \\ &= Ae(t) + (A - BK)\tilde{x}(t_k). \end{aligned} \tag{11.15}$$

By solving the differential equation (11.15), we have

$$e(t) = e^{A(t-t_k)}e(t_k) + \int_{t_k}^{t} e^{A(t-s)}(A - BK)\tilde{x}(t_k)\mathrm{d}s, \ t \in [t_k, t_{k+1})$$

which further implies that

$$\|e(t)\| \leq e^{\|A\|(t-t_k)}\|e(t_k)\| + \int_{t_k}^{t} e^{\|A\|(t-s)}\|A - BK\|\|\tilde{x}(t_k)\|\mathrm{d}s. \tag{11.16}$$

According to the definition of the updating instants (11.3), the following statements are true for $k \in \mathbb{N}$,

$$\begin{cases} \|e(t_k)\| = \|x(t_k) - \tilde{x}(t_k)\| < \alpha \\ \|e(t_{k+1}^-)\| = \|x(t_{k+1}) - \tilde{x}(t_k)\| = \alpha. \end{cases}$$

In the following, four cases will be discussed in details. If $\|\tilde{x}(t_k)\| = 0$, and $\|A\| = 0$, which implies that $\|e(t)\| \leq \|e(t_k)\|$ always holds for $k \in \mathbb{N}$, i.e., no event will occur, thus $t_{\min}^l > 0$.

If $\|\tilde{x}(t_k)\| = 0$ and $\|A\| \neq 0$, we get $\|e(t)\| \leq e^{\|A\|(t-t_k)}\|e(t_k)\|$, thus, solving the equality $e^{\|A\|(t-t_k)}\|e(t_k)\| = \alpha$ yields that $t \triangleq t_{k+1}' = t_k + \frac{\ln \alpha - \ln \|e(t_k)\|}{\|A\|}$. It is obviously that $t_{k+1}' \leq t_{k+1}$ according to (11.3). Thus, $t_{k+1}' - t_k \leq t_{k+1} - t_k$ for $\forall k \in \mathbb{N}$, and the following inequality always holds $t_{k+1}' - t_k = \frac{\ln \alpha - \ln \|e(t_k)\|}{\|A\|} > 0$, $\forall k \in \mathbb{N}$, which indicates $t_{\min}^l > 0$.

If $\|\tilde{x}(t_k)\| \neq 0$, and $\|A\| = 0$, we have $\|e(t)\| \leq \|e(t_k)\| + \|A - BK\|\|\tilde{x}(t_k)\|(t - t_k)$. Similarly, we have $t_{k+1}' - t_k = \frac{\alpha - \|e(t_k)\|}{\|A - BK\|\|\tilde{x}(t_k)\|} > 0$, $\forall k \in \mathbb{N}$, which means $t_{\min}^l > 0$.

If $\|\tilde{x}(t_k)\| \neq 0$ and $\|A\| \neq 0$, we have $\|e(t)\| \leq \left(\|e(t_k)\| + \frac{\|A-BK\|\|\tilde{x}(t_k)\|}{\|A\|}\right) e^{\|A\|(t-t_k)} - \frac{\|A-BK\|\|\tilde{x}(t_k)\|}{\|A\|}$. Similarly, we have $t'_{k+1} - t_k = \frac{1}{\|A\|} \ln\left(1 + \frac{\alpha - \|e(t_k)\|}{\|A-BK\|\|\tilde{x}(t_k)\|}\|A\|\right) > 0$, $\forall k \in \mathbb{N}$, which also indicates $t^l_{\min} > 0$.

Therefore, the Zeno behavior can be excluded since there always exists $t^l_{\min} > 0$ such that $t_{\min} \geq t^l_{\min} > 0$. □

11.4 QUANTIZED EVENT-TRIGGERED CONTROL: RELATIVE THRESHOLD STRATEGY

In the previous section, the event-triggered control scheme with fixed event threshold is proposed. In this section, the event-triggering function and the quantization length will be modified by using the information of quantized state and the asymptotic stability of the quantized event-triggered control systems is also guaranteed.

First, the quantization length is modified as

$$\Delta(t_{k+1}) = \eta\|\tilde{x}(t_k)\| \tag{11.17}$$

where $0 < \eta < 1$. It is clear that the modified quantization length is dependent on the quantized state, which implies that the quantization length is varying according the quantized state accordingly.

By simply replacing Δ with $\Delta(t_{k+1})$ in Lemma 11.1, we get the following similar results.

Lemma 11.3. *There is only one element in $\Delta(t_{k+1})M(i)\mathbf{1}$ is not equal to 0 and $\|\Delta(t_{k+1})M(i)\mathbf{1}\| = \Delta(t_{k+1})$.*

Lemma 11.4. *With the definition of quantization length (11.17), if $\tilde{x}(t_0) \neq 0$, the quantized state $\tilde{x}(t_k) \neq 0$ holds for $\forall k \in \mathbb{N}$.*

Proof. The desired results are shown by the contradiction. Without loss of generality, suppose $\tilde{x}(t_{l+1}) = 0$ and $\tilde{x}(t_l) \neq 0$ for some $l \in \mathbb{N}$. Similar to (11.8) and (11.9), we have $\tilde{x}(t_{l+1}) = \Delta(t_{l+1})M(q_{l+1})\mathbf{1} + \tilde{x}(t_l)$. Then, according to Lemma 11.3, we obtain the following inequality since $0 < \eta < 1$, $\Delta(t_{l+1}) = \|\tilde{x}(t_{l+1}) - \tilde{x}(t_l)\| = \|\tilde{x}(t_l)\| > \eta\|\tilde{x}(t_l)\|$, which is in contradiction to (11.17). Thus, we have $\tilde{x}(t_k) \neq 0$, $\forall k \in \mathbb{N}$. □

Now, the modified event-triggering function $\bar{\zeta}(t)$ relating to the quantized states $\tilde{x}(t_k)$ can be defined as:

$$\bar{\zeta}(t) = \frac{\|e(t)\|}{\gamma\|\tilde{x}(t_k)\|} = \frac{\|x(t) - \tilde{x}(t_k)\|}{\gamma\|\tilde{x}(t_k)\|}, \; t \in [t_k, t_{k+1}) \tag{11.18}$$

where γ satisfies that $0 < \gamma\alpha < 1$. Combining (11.18) with (11.3) gives that $\|x(t) - \tilde{x}(t_k)\| < \gamma\alpha\|\tilde{x}(t_k)\|$, which is satisfied before the next updating instant t_{k+1}, also we have $\|x(t_k) - \tilde{x}(t_k)\| < \gamma\alpha\|\tilde{x}(t_k)\|$. The quantized state between two consecutive updating instants satisfies

$$\begin{aligned}
\tilde{x}(t_{k+1}) &= \tilde{e}(t^-_{k+1}) + \tilde{x}(t_k) \\
&= \Delta(t_{k+1})M(q_{k+1})\mathbf{1} + \tilde{x}(t_k) \\
&= \eta\|\tilde{x}(t_k)\|M(q_{k+1})\mathbf{1} + \tilde{x}(t_k), k \in \mathbb{N}.
\end{aligned} \tag{11.19}$$

Similarly, we have the following results.

Lemma 11.5. *The inequality $|e_{\psi(q_{k+1})}(t_{k+1}^-)| \geq \frac{\gamma\alpha}{\sqrt{n}}\|\tilde{x}(t_k)\|$ holds.*

Theorem 11.4. *For the quantization length (11.17), if η is chosen to satisfy the following inequalities:*

$$\begin{cases} \eta \leq \frac{2}{\sqrt{n}} \frac{\gamma\alpha - \sqrt{n}\gamma^2\alpha^2}{1 - \gamma^2\alpha^2} \\ 0 < \gamma\alpha < \frac{1}{\sqrt{n}} \end{cases} \tag{11.20}$$

where γ and α are the parameters defined above, then $e(t)$ will never exceed the event threshold $\gamma\alpha\|\tilde{x}(t_{k+1})\|$ for $t \in [t_k, t_{k+1})$.

Proof. By the definition of the UIS (11.3) and the modified event-triggering function $\bar{\zeta}(t)$ in (11.18), we have $\|x(t_{k+1}) - \tilde{x}(t_{k+1})\| < \gamma\alpha\|\tilde{x}(t_{k+1})\|$, which together with (11.17) and (11.19) yields that

$$\begin{aligned} \|x(t_{k+1}) - \tilde{x}(t_{k+1})\| &= \|e(t_{k+1}^-) - \tilde{e}(t_{k+1})\| \\ &= \|e(t_{k+1}^-) - \Delta(t_{k+1})M(q_{k+1})\mathbf{1}\| \\ &= (\gamma^2\alpha^2\|\tilde{x}(t_k)\|^2 + \eta^2\|\tilde{x}(t_k)\|^2 - 2\eta\|\tilde{x}(t_k)\|\|e_{\psi(q_{k+1})}(t_{k+1}^-)\|)^{\frac{1}{2}} \\ &< \gamma\alpha\|\tilde{x}(t_{k+1})\|. \end{aligned} \tag{11.21}$$

With the modified quantization length (11.17), it is true that

$$\begin{aligned} \|\tilde{x}(t_{k+1})\|^2 &= \|\tilde{x}(t_{k+1}) - \tilde{x}(t_k) + \tilde{x}(t_k)\|^2 \\ &\geq \|\|\tilde{x}(t_{k+1}) - \tilde{x}(t_k)\| - \|\tilde{x}(t_k)\|\|^2 \\ &= (\eta - 1)^2\|\tilde{x}(t_k)\|^2. \end{aligned} \tag{11.22}$$

It is clear that if η is chosen such that

$$0 < \eta < \frac{2|e_{\psi(q_{k+1})}(t_{k+1}^-)|/\|\tilde{x}(t_k)\| - 2\gamma^2\alpha^2}{1 - \gamma^2\alpha^2}$$

which implies that

$$\gamma^2\alpha^2\|\tilde{x}(t_k)\|^2 + \eta^2\|\tilde{x}(t_k)\|^2 - 2\eta\|\tilde{x}(t_k)\|\|e_{\psi(q_{k+1})}(t_{k+1}^-)\| \leq \gamma^2\alpha^2(\eta - 1)^2\|\tilde{x}(t_k)\|^2,$$

the inequality (11.21) can be guaranteed.

Furthermore, according to Lemma 11.5, we have $|e_{\psi(q_{k+1})}(t_{k+1}^-)| \geq \frac{\gamma\alpha}{\sqrt{n}}\|\tilde{x}(t_k)\|$, thus η can be selected to satisfy that $\eta < \frac{2}{\sqrt{n}}\frac{\gamma\alpha - \sqrt{n}\gamma^2\alpha^2}{1 - \gamma^2\alpha^2}$, and $0 < \gamma\alpha < \frac{1}{\sqrt{n}}$ should be also satisfied to make sure $\eta > 0$, thus the proof is completed. □

Theorem 11.5. *Consider the closed-loop system (11.10) and (11.3) with the modified event-triggering function $\bar{\zeta}(t)$ in (11.18). If there exists matrix $P > 0$ satisfying the Riccati equation (11.11) for given $Q > 0$, and γ is selected such that the following inequality holds,*

$$\gamma\alpha < \min\left\{\frac{\sqrt{\lambda_{\min}(Q)}}{\frac{1}{2}\|PBB^TP\| + \sqrt{\lambda_{\min}(Q)}}, \frac{1}{\sqrt{n}}\right\} \tag{11.23}$$

then the controller gain can be designed as $K = -\frac{1}{2}B^TP$ and the states of closed-loop system (11.10) are globally asymptotically stable.

Proof. The inequality (11.12) can be rewritten as $\dot{V}(t) \leq -\lambda_{\min}(Q)\|x(t)\|^2 + \gamma^2\|P\bar{B}\|^2\|\tilde{x}(t_k)\|^2\bar{\zeta}^2(t)$. If $\|x(t)\| \geq \|\tilde{x}(t_k)\|$, we have

$$\dot{V}(t) \leq -\lambda_{\min}(Q)\|x(t)\|^2 + \gamma^2\mu^2\|\tilde{x}(t_k)\|^2$$
$$\leq -\lambda_{\min}(Q)\|x(t)\|^2 + \gamma^2\mu^2\|x(t)\|^2$$
$$\leq \frac{-\lambda_{\min}(Q) + \gamma^2\mu^2}{\lambda_{\min}(P)}x^T(t)Px(t)$$
$$= \frac{-\lambda_{\min}(Q) + \gamma^2\mu^2}{\lambda_{\min}(P)}V(t).$$

Thus, if $\gamma\alpha < \frac{\sqrt{\lambda_{\min}(Q)}}{\frac{1}{2}\|PBB^TP\|}$, we always have $V(t) \leq e^{-\phi(t-t_0)}V(t_0)$, $t \in [t_0, \infty)$, where $\phi = \frac{\lambda_{\min}(Q)-\gamma^2\mu^2}{\lambda_{\min}(P)}$.

If $\|x(t)\| < \|\tilde{x}(t_k)\|$, according to (11.3) and the following norm inequality,

$$\left|\|x(t)\| - \|\tilde{x}(t_k)\|\right| = \|\tilde{x}(t_k)\| - \|x(t)\|$$
$$\leq \|x(t) - \tilde{x}(t_k)\| \leq \gamma\alpha\|\tilde{x}(t_k)\|$$

we obtain that $\|\tilde{x}(t_k)\| \leq \frac{1}{1-\gamma\alpha}\|x(t)\|$.

Then, we get

$$\dot{V}(t) \leq -\lambda_{\min}(Q)\|x(t)\|^2 + \gamma^2\mu^2\|\tilde{x}(t_k)\|^2$$
$$\leq -\lambda_{\min}(Q)\|x(t)\|^2 + (\frac{\gamma\mu}{1-\gamma\alpha})^2\|x(t)\|^2$$
$$\leq \frac{-\lambda_{\min}(Q) + (\frac{\gamma\mu}{1-\gamma\alpha})^2}{\lambda_{\min}(P)}x^T(t)Px(t)$$
$$= \frac{-\lambda_{\min}(Q) + (\frac{\gamma\mu}{1-\gamma\alpha})^2}{\lambda_{\min}(P)}V(t).$$

Therefore, if $\gamma\alpha < \frac{\sqrt{\lambda_{\min}(Q)}}{\frac{1}{2}\|PBB^TP\| + \sqrt{\lambda_{\min}(Q)}}$, we have $V(t) \leq e^{-\theta(t-t_0)}V(t_0)$, $t \in [t_0, \infty)$, where $\theta = \frac{\lambda_{\min}(Q)-(\frac{\gamma\mu}{1-\gamma\alpha})^2}{\lambda_{\min}(P)}$.

Considering (11.20), if (11.23) is satisfied, we always have $\lim_{t\to+\infty} V(t) = 0$, which further implies that the quantized event-triggered control system is asymptotically stable, thus the proof is completed. □

Theorem 11.6. *For the updating instants defined by (11.3) with the modified event-triggering function $\bar{\zeta}(t)$ in (11.18), there exists a strictly positive lower bound $t^l_{\min}$ such that $t_{\min} \geq t^l_{\min} > 0$.*

Proof. Following a similar line of the proof of Theorem 11.3, it is easy to show that the inequality (11.16) still holds, and the following statements are true for $k \in \mathbb{N}$,

$$\begin{cases} \|e(t_k)\| = \|x(t_k) - \tilde{x}(t_k)\| < \gamma\alpha\|\tilde{x}(t_k)\| \\ \|e(t_{k+1}^-)\| = \|x(t_{k+1}) - \tilde{x}(t_k)\| = \gamma\alpha\|\tilde{x}(t_k)\|. \end{cases}$$

If $\|A\| = 0$, we have $t^l_{k+1} - t_k \geq \frac{\gamma\alpha\|\tilde{x}(t_k)\| - \|e(t_k)\|}{\|A - BK\|\|\tilde{x}(t_k)\|} > 0$, which means $t^l_{min} > 0$. Otherwise, if $\|A\| \neq 0$, similarly, we have $e^{\|A\|(t'_{k+1} - t_k)} = 1 + \frac{\gamma\alpha\|\tilde{x}(t_k)\| - \|e(t_k)\|}{\|A - BK\|\|\tilde{x}(t_k)\|}\|A\| > 0$, which also indicates $t^l_{min} > 0$. Therefore, the Zeno behavior can be excluded since $t^l_{min} > 0$. □

11.5 NUMERICAL SIMULATION

Consider the following simplified model for a planar vertical take-off and landing (PVTOL) aircraft shown in Figure 11.4,

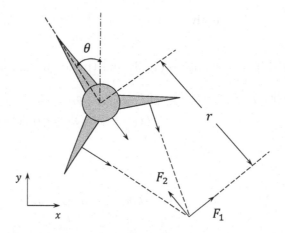

Figure 11.4 Schematic model of PVTOL aircraft

$$
\begin{bmatrix} \dot{x} \\ \dot{y} \\ \dot{\theta} \\ \ddot{x} \\ \ddot{y} \\ \ddot{\theta} \end{bmatrix} = \begin{bmatrix} 0 & 0 & 0 & 1 & 0 & 0 \\ 0 & 0 & 0 & 0 & 1 & 0 \\ 0 & 0 & 0 & 0 & 0 & 1 \\ 0 & 0 & -g & -c/m & 0 & 0 \\ 0 & 0 & 0 & 0 & -c/m & 0 \\ 0 & 0 & 0 & 0 & 0 & 0 \end{bmatrix} \begin{bmatrix} x \\ y \\ \theta \\ \dot{x} \\ \dot{y} \\ \dot{\theta} \end{bmatrix} + \begin{bmatrix} 0 & 0 \\ 0 & 0 \\ 0 & 0 \\ 1/m & 0 \\ 0 & 1/m \\ r/J & 0 \end{bmatrix} \begin{bmatrix} u_1 \\ u_2 \end{bmatrix}
$$

where x, y, and θ denote the position and orientation of the mass center of the aircraft, m, J, r, g and c, respectively, represent the mass of the vehicle, the moment of inertia, thrust offset, the gravitational constant, and the rotational damping, F_1 and F_2 are a pair of forces of the main downward thruster and the maneuvering thrusters, $u_1 = F_1$ and $u_2 = F_2 - mg$. We assume that $m = 4$ kg, $J = 0.0475$ kgm^2, $r = 0.25$ m, $g = 9.8$ m/s and $c = 0.05$ Ns/m. The initial conditions are $x_0 = 4$, $y_0 = 3$ and $\theta_0 = 0.2$.

Setting $Q = 100I_6$ and solving the Riccati equation (11.11), we can obtain the state feedback controller gain

$$
K = \frac{1}{2}B^T P = \begin{bmatrix} 5.0249 & -0.0000 & -29.3124 & 7.4048 & -0.0000 & -5.9044 \\ 0.0000 & -5.0249 & -0.0000 & 0.0000 & -6.7093 & -0.0000 \end{bmatrix}.
$$

We first examine the validity of the quantized event-triggered control strategy for the fixed event threshold case. Taking $\alpha = 0.3$ and $\Delta = 0.15$, and applying the proposed quantized event-triggered controller to the plant, the simulation results are shown in Figure 11.5. Figure 11.5(a–c) depicts the state $x(t)$ and its quantized value and Figure 11.5(d) illustrates the signal $\|e(t)\|$ and inter-event intervals. It is easy to find that the states of the closed-loop system are globally UUB and the proposed control approach is effective.

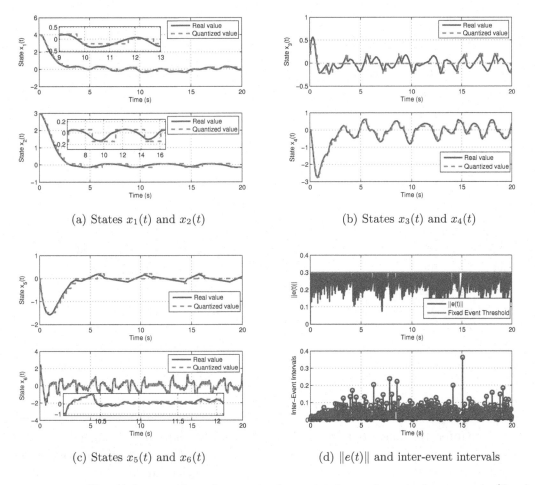

(a) States $x_1(t)$ and $x_2(t)$

(b) States $x_3(t)$ and $x_4(t)$

(c) States $x_5(t)$ and $x_6(t)$

(d) $\|e(t)\|$ and inter-event intervals

Figure 11.5 Simulation results of quantized event-triggered control approach (fixed event threshold case)

Next, we will show the validity of the quantized event-triggered control strategy for the relative event threshold case. By selecting $\gamma\alpha = 0.3$ and $\eta = 0.06$, the simulation results are shown in Figure 11.6. Figure 11.6(a–c) presents the state $x(t)$ and its quantized value, and Figure 11.6(d) depicts the signal $\|e(t)\|$ and inter-event intervals. It is clear that the states of the closed-loop system are globally asymptotically stable, which verifies the effectiveness of the proposed control approach.

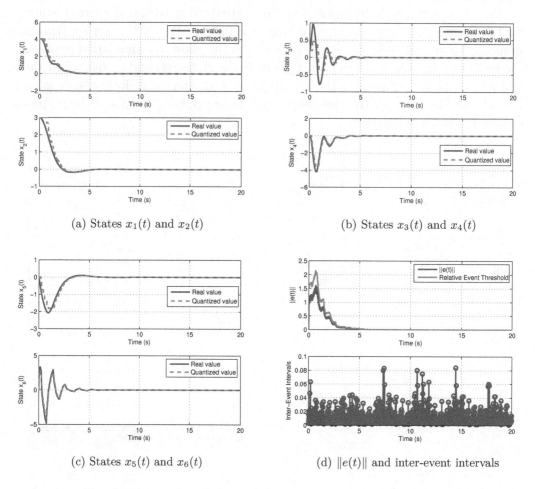

(a) States $x_1(t)$ and $x_2(t)$ (b) States $x_3(t)$ and $x_4(t)$

(c) States $x_5(t)$ and $x_6(t)$ (d) $\|e(t)\|$ and inter-event intervals

Figure 11.6 Simulation results of quantized event-triggered control approach (relative event threshold case)

11.6 SUMMARY

In this chapter, the quantized event-triggered controllers have been designed for linear systems by using event-triggered quantizers. For both the fixed and relative event threshold cases, the upper bounds of the quantization length have been calculated, and the stability analysis of the closed-loop systems have been performed. It has been shown that there always exists a lower bound of the minimum inter-event interval. Finally, a numerical example has been presented to show the validity of the proposed approaches.

Event-Triggered Disturbance Rejection Control for CPSs

12.1 INTRODUCTION

It is well-known that disturbances widely exist in practical control systems and bring adverse effects on the performance of control systems. The active disturbance rejection control (ADRC) scheme provides an effective way to achieve the disturbance rejection and enhance the system robustness. During the past several decades, the ADRC schemes have been extensively investigated [59, 54, 144, 145, 230]. More recently, some event-triggered ADRC approaches are proposed to save the communication resource while maintaining desired closed-loop system performance and disturbance rejection ability. In [134], the event-triggered ADRC approach is proposed to achieve position tracking of DC torque motors, where an event-based sampler together with an extended state observer is introduced to jointly observe the system state. In [141], by using disturbance/uncertainty estimation and attenuation techniques, the sampled-data-based event-triggered ADRC approach is proposed for disturbed systems in networked environment, where only measurable outputs are necessary for controller design, the communication frequency is remarkably reduced and the closed-loop system performance can also be maintained at a satisfactory level. It should be pointed out that, the aforementioned results are developed for continuous-time controlled plant, how to design the effective event-triggered controllers for discrete-time systems subject to external disturbance undoubtedly poses an interesting and important problem both in theory and practice.

This chapter proposes the event-triggered control approaches for discrete-time systems with external disturbance. Due to the existence of unknown external disturbance, it is imprecise to directly use the system states to design the UIS and feedback controllers. To address this problem, the states and disturbance are estimated by using an extended state observer (ESO) and then these estimates are utilized to design the UIS and controller. Since in many systems, at least one state variable can be either measured directly or calculated easily from the output, the reduced-order ESO and extended functional state observer (EFSO) cases are further considered. Furthermore, the stability and disturbance rejection properties are analyzed. Numerical and experimental results are given to demonstrate the validity and potential of the proposed event-triggered control strategies.

12.2   SYSTEM DESCRIPTION

In this chapter, we focus on designing the event-triggered disturbance rejection control strategy for discrete-time linear systems with disturbance. The controlled plant is a discrete-time system given by

$$
\begin{aligned}
x(k+1) &= Ax(k) + Bu(k) + B_d d(k), \\
y(k) &= Cx(k),
\end{aligned}
\tag{12.1}
$$

where $u(k) \in \mathbb{R}^m$ denotes the control input, $x(k) \in \mathbb{R}^n$ denotes the system state, $y(k) \in \mathbb{R}^q$ denotes the system output, and $d(k) \in \mathbb{R}^p$ denotes the unknown external disturbance. A, B, B_d, and C are known system matrices, and $\text{rank}(B) = m$, $\text{rank}(B_d) = p$, and $\text{rank}(C) = q$. In this chapter, some necessary assumptions are given as follows:

Assumption 12.1. *Both the pairs (A, B) and (A, C) are stabilizable and detectable, respectively, and matrices A, B_d, and C satisfy that*

$$
rank \left(\begin{bmatrix} I_n - A & B_d \\ C & 0 \end{bmatrix} \right) = n + p.
\tag{12.2}
$$

Assumption 12.2. *The triple (A, B, C) is invertible, and no invariant zeros located at $z = 1$ for (A, B, C).*

Assumption 12.3. *The external disturbance $d(k) \in \mathbb{R}^p$ is unknown but bounded, and $h(k) \triangleq d(k+1) - 2d(k) + d(k-1)$ is also bounded, i.e., $\|d(k)\| \leq \bar{d}$ and $\|h(k)\| \leq \delta_d$.*

Remark 12.1. *Provided that the discrete-time system (12.1) is discretized from its continuous counterpart with the sampling period T_s,*

$$
\begin{aligned}
\dot{x}(t) &= G_c x(t) + H_c u(t) + H_d d(t) \\
y(t) &= Cx(t).
\end{aligned}
$$

According to the ZOH properties, it can be obtained that $u(k) = u(kT_s)$, $x(k) = x(kT_s)$, $y(k) = y(kT_s)$, $d(k) = d(kT_s)$, and $A = e^{G_c T_s}$, $B = \int_0^{T_s} e^{G_c \tau} H_c d\tau$, $B_d = \int_0^{T_s} e^{G_c \tau} H_d d\tau$, and $d(k) = \int_0^{T_s} e^{G_c \tau} d((k+1)T_s - \tau) d\tau$ can be calculated. Moreover, if $d(t)$ is smooth and bounded, then $\delta_d \in O(T_s^3)$ always holds [1].

Because of the control performance is always affected by the unknown disturbance $d(k)$ in (12.1), it is important to actively reject the negative effects induced by the disturbance. To estimate the unknown disturbance, we first define the extended state as $\bar{x}(k) = [x^T(k)\ d^T(k-1)\ d^T(k)]^T$ and system (12.1) is extended as follows:

$$
\begin{aligned}
\bar{x}(k+1) &= \bar{A}\bar{x}(k) + \bar{B}u(k) + Eh(k), \\
y(k) &= \bar{C}\bar{x}(k),
\end{aligned}
\tag{12.3}
$$

where

$$\bar{A} = \begin{bmatrix} A & 0 & B_d \\ 0 & 0 & I_p \\ 0 & -I_p & 2I_p \end{bmatrix}, \ \bar{B} = \begin{bmatrix} B \\ 0 \\ 0 \end{bmatrix}, \ \bar{C} = \begin{bmatrix} C & 0 & 0 \end{bmatrix}, \ E = \begin{bmatrix} 0 & 0 & I_p \end{bmatrix}^T.$$

In this chapter, the problem of event-triggered disturbance rejection control will be addressed for discrete-time systems with unknown disturbance. More specifically, we are interested in designing the full-order ESO-, reduced-order ESO- and EFSO-based event-triggered disturbance rejection controllers and performing the stability and disturbance rejection performance analysis.

12.3 FULL-ORDER ESO-BASED EVENT-TRIGGERED DISTURBANCE RE-JECTION CONTROL

In this section, the following full-order ESO can be designed for system (12.3),

$$\hat{\bar{x}}(k+1) = \bar{A}\hat{\bar{x}}(k) + \bar{B}u(k) + L(y(k) - \bar{C}\hat{\bar{x}}(k)), \tag{12.4}$$

where L is the observer gain with appropriate dimension and $\hat{\bar{x}}(k)$ is the estimated state.

Remark 12.2. *With the state estimate $\hat{\bar{x}}(k)$, both the disturbance estimate $\hat{d}(k)$ and the state estimate $\hat{x}(k)$ can be obtained simultaneously. Then, $\hat{d}(k)$ will be incorporated in the controller to achieve active disturbance rejection.*

To reduce the communication frequency in the CPSs, the event-triggered control method is always implemented. In this chapter, the discrete-time event-triggered controller is designed by utilizing the estimated state $\hat{\bar{x}}(k_i)$, where k_i, $i = 0, 1, 2, \ldots$ are the updating instants of the controller. Meanwhile, a ZOH is deployed to hold the control signals during $[k_i, k_{i+1})$. More specifically, we design the event-triggered controller as

$$u(k) = K\hat{\bar{x}}(k_i) = K_x\hat{x}(k_i) + K_d\hat{d}(k_i), \ k \in [k_i, k_{i+1}), \tag{12.5}$$

where K_x denotes the feedback control gain and K_d denotes the disturbance compensation gain, and define the controller gain as $K = \begin{bmatrix} K_x & 0 & K_d \end{bmatrix}$. For simplicity, we assume that the first event occurs at $k_0 = 0$.

Remark 12.3. *From controller (12.5), the control input is updated at instant k_i, and both the disturbance estimate $\hat{d}(k_i)$ and the state estimate $\hat{x}(k_i)$ will be received by the controller, which are held during $[k_i, k_{i+1})$, and thus the controller updating frequency will be reduced substantially.*

It is worth mentioning that the updating instants k_i, $i = 0, 1, 2, \ldots$, of the controller should be predefined based upon the physical available variables. With the estimated state $\hat{\bar{x}}(k)$, the "event" is defined by using the error $e(k) = \hat{\bar{x}}(k) - \hat{\bar{x}}(k_i)$. Thus, the updating instants k_i, $i = 1, 2, \ldots$, are determined by

$$k_{i+1} = \sup \left\{ k > k_i \left| e^T(k)\Theta e(k) \le \varepsilon + \varrho\hat{\bar{x}}^T(k)\Theta\hat{\bar{x}}(k) \right. \right\}, \tag{12.6}$$

where ε and ϱ are the user-defined parameters and Θ is a weighting matrix to be determined. Note that $e(k_i) \equiv 0$ always holds at each updating instant k_i.

Remark 12.4. *According to [106], the event is generated whenever the state $x(t)$ leaves the surrounding $\Omega = \{x(t) \mid \|e(t)\|^2 \leq \varepsilon\}$, which implies that the updating instants can be obtained from*

$$t_{k+1} = \inf \left\{ t > t_k \, \middle| \, \|e(t)\|^2 > \varepsilon \right\}. \tag{12.7}$$

For the discrete-time systems considered in this chapter, it is straightforward to define the following updating instants as the discrete-time counterpart of (12.7):

$$k_{i+1} = \inf \left\{ k > k_i \mid \|e(k)\|^2 > \varepsilon \right\}, \quad k \in \mathbb{N}, \tag{12.8}$$

which is quite different from the updating instants defined by (12.6). It is clear that, according to (12.6), the updating instant is the maximum integer k satisfying $\|e(k)\|^2 \leq \varepsilon + \varrho\hat{\bar{x}}^T(k)\Theta\hat{\bar{x}}(k)$. However, if we take the similar mechanism as (12.8), the $e^T(k)\Theta e(k)$ may larger $\varepsilon + \varrho\hat{\bar{x}}^T(k)\Theta\hat{\bar{x}}(k)$ at the updating instant, which may lead to unprecise derivation of the stability analysis. Furthermore, variable $e(k+1)$ can be calculated by $e(k+1) = \hat{\bar{x}}(k+1) - \hat{\bar{x}}(k_i)$ and $\hat{\bar{x}}(k+1)$ can be obtained by equation (12.4).

Remark 12.5. *It is easy to see that, according to the definition of updating instants (12.6), the following two facts are true, (1) the UIS (12.6) includes the special case $k_{i+1} = k_i + 1$ in some conditions, which indicates that the conventional periodic updating way is contained in the proposed event-triggered updating schemes and (2) in the time interval $[k_i, k_{i+1})$, the inequality $e^T(k)\Theta e(k) \leq \varepsilon + \varrho\hat{\bar{x}}^T(k)\Theta\hat{\bar{x}}(k)$ always holds, which will be used in the stability analysis.*

Define the error signal $e(k)$ as

$$e(k) = \hat{\bar{x}}(k) - \hat{\bar{x}}(k_i) = \begin{bmatrix} e_1(k) \\ e_2(k) \\ e_3(k) \end{bmatrix} = \begin{bmatrix} \hat{x}(k) - \hat{x}(k_i) \\ \hat{d}(k-1) - \hat{d}(k_i - 1) \\ \hat{d}(k) - \hat{d}(k_i) \end{bmatrix}.$$

For $k \in [k_i, k_{i+1})$, the closed-loop system can be determined by substituting the event-triggered controller (12.5) into system (12.1) as

$$\begin{aligned} x(k+1) &= Ax(k) + B(K_x\hat{x}(k_i) + K_d\hat{d}(k_i)) + B_d d(k) \\ &= (A + BK_x)x(k) - BK\tilde{\bar{x}}(k) - BKe(k) + (B_d + BK_d)d(k). \end{aligned} \tag{12.9}$$

Moreover, for the ESO (12.4), the estimation error system can be given as

$$\tilde{\bar{x}}(k+1) = (\bar{A} - L\bar{C})\tilde{\bar{x}}(k) + Eh(k), \tag{12.10}$$

where $\tilde{\bar{x}}(k) = \bar{x}(k) - \hat{\bar{x}}(k)$ is the estimation error.

Define $\xi(k) = [\tilde{x}^T(k), x^T(k)]^T$ as an extended state variable. Based on systems (12.9) and (12.10), the closed-loop system is written as

$$\xi(k+1) = \tilde{A}\xi(k) + \tilde{B}_1 e(k) + \tilde{B}_2 f(k), k \in [k_i, k_{i+1}), \tag{12.11}$$

where $f(k) = [d^T(k), h^T(k)]^T$,

$$\tilde{A} = \begin{bmatrix} \bar{A} - L\bar{C} & 0 \\ -BK & A + BK_x \end{bmatrix}, \quad \tilde{B}_1 = \begin{bmatrix} 0 \\ -BK \end{bmatrix}, \quad \tilde{B}_2 = \begin{bmatrix} 0 & E \\ B_d + BK_d & 0 \end{bmatrix}.$$

Meanwhile, K_x and L should be selected to guarantee that $A + BK_x$ and $\bar{A} - L\bar{C}$ are Schur, respectively.

Lemma 12.1. *Under the Assumption 12.1, the pair $(\bar{A}, \bar{C})$ is detectable.*

Proof. From the well-known linear systems theory [164], the pair (A, C) is detectable if and only if $\operatorname{rank} \begin{bmatrix} zI_n - A \\ C \end{bmatrix} = n, \forall z \in \mathbb{C}, |z| \geq 1$. Then, we have

$$\operatorname{rank} \begin{bmatrix} zI_{n+2p} - \bar{A} \\ \bar{C} \end{bmatrix} = \operatorname{rank} \begin{bmatrix} zI_n - A & 0 & -B_d \\ 0 & zI_p & -I_p \\ 0 & I_p & (z-2)I_p \\ C & 0 & 0 \end{bmatrix}. \tag{12.12}$$

Case I: When $z \neq 1$, equation (12.12) can be rewritten as

$$M \begin{bmatrix} zI_{n+2p} - \bar{A} \\ \bar{C} \end{bmatrix} N = \begin{bmatrix} zI_n - A & -B_d & -B_d \\ 0 & (z-1)I_p & -I_p \\ 0 & 0 & (z-1)I_p \\ C & 0 & 0 \end{bmatrix},$$

where

$$M = \begin{bmatrix} I_n & 0 & 0 & 0 \\ 0 & I_p & 0 & 0 \\ 0 & -I_p & I_p & 0 \\ 0 & 0 & 0 & I_q \end{bmatrix}, \quad N = \begin{bmatrix} I_n & 0 & 0 \\ 0 & I_p & 0 \\ 0 & I_p & I_p \end{bmatrix},$$

which is equivalent to

$$\operatorname{rank} \begin{bmatrix} zI_{n+p} - \bar{A} \\ \bar{C} \end{bmatrix} = \operatorname{rank} \begin{bmatrix} zI_n - A \\ C \end{bmatrix} + 2p = n + 2p.$$

Case II: When $z = 1$, under Assumption 12.1, we have

$$\operatorname{rank} \begin{bmatrix} I_{n+p} - \bar{A} \\ \bar{C} \end{bmatrix} = \operatorname{rank} \begin{bmatrix} I_n - A & 0 & -B_d \\ 0 & I_p & -I_p \\ C & 0 & 0 \end{bmatrix}$$

$$= \operatorname{rank} \begin{bmatrix} I_n - A & -B_d \\ C & 0 \end{bmatrix} + p = n + 2p.$$

Thus, we always have rank $\begin{bmatrix} zI_{n+p} - \bar{A} \\ \bar{C} \end{bmatrix} = n + 2p$, that is, the pair $(\bar{A}, \bar{C})$ is detectable. This completes the proof. □

Now, we will show that states of system (12.11) are UUB.

Theorem 12.1. *For system (12.11), if there exists $P > 0$ satisfying*

$$Q = 3\tilde{A}^T P \tilde{A} - P + 3\varrho(\Lambda_{11} + I_{2(n+p)}) \leq 0 \tag{12.13}$$

where $\Lambda_{11} = \begin{bmatrix} \tilde{B}_1^T P \tilde{B}_1 & -\tilde{B}_1^T P \tilde{B}_{11} \\ -\tilde{B}_{11}^T P \tilde{B}_1 & \tilde{B}_{11}^T P \tilde{B}_{11} \end{bmatrix}$ *with* $\tilde{B}_1 = [\tilde{B}_{11} \ \tilde{B}_{12}]$, $\tilde{B}_{11} = \begin{bmatrix} 0 \\ -BK_x \end{bmatrix}$ *and*

$\tilde{B}_{12} = \begin{bmatrix} 0 & 0 \\ 0 & -BK_d \end{bmatrix}$, *then the states of system (12.11) are UUB and the weighting matrix in (12.6) can be designed as* $\Theta = \tilde{B}_1^T P \tilde{B}_1$.

Proof. According to inequality (12.13), we obtain $-Q = P - 3\tilde{A}^T P \tilde{A} - 3\varrho(\Lambda_{11} + I_{2(n+p)})$. Since $3\tilde{A}^T P \tilde{A} \geq 0$ and $3\varrho(\Lambda_{11} + I_{2(n+p)}) > 0$ hold, it can be verified that $\delta \triangleq \frac{\lambda_{\min}(-Q)}{\lambda_{\max}(P)} < 1$. Consider the Lyapunov function as $V(k) = \xi^T(k) P \xi(k)$, and the increment of $V(k)$ for $k \in [k_i, k_{i+1})$ is obtained as

$$\begin{aligned}
\Delta V(k) &= V(k+1) - V(k) \\
&= \xi^T(k)(\tilde{A}^T P \tilde{A} - P)\xi(k) + \xi^T(k)\tilde{A}^T P \tilde{B}_1 e(k) + \xi^T(k)\tilde{A}^T P \tilde{B}_2 f(k) \\
&\quad + e^T(k)\tilde{B}_1^T P \tilde{A}\xi(k) + e^T(k)\tilde{B}_1^T P \tilde{B}_1 e(k) \\
&\quad + e^T(k)\tilde{B}_1^T P \tilde{B}_2 f(k) + f^T(k)\tilde{B}_2^T P \tilde{A}\xi(k) \\
&\quad + f^T(k)\tilde{B}_2^T P \tilde{B}_1 e(k) + f^T(k)\tilde{B}_2^T P \tilde{B}_2 f(k) \\
&\leq \xi^T(k)(3\tilde{A}^T P \tilde{A} - P)\xi(k) + 3e^T(k)\tilde{B}_1^T P \tilde{B}_1 e(k) + 3f^T(k)\tilde{B}_2^T P \tilde{B}_2 f(k).
\end{aligned}$$

Since $e^T(k)\Theta e(k) \leq \varepsilon + \varrho \hat{\bar{x}}^T(k)\Theta \hat{\bar{x}}(k)$ and the weighting matrix $\Theta = \tilde{B}_1^T P \tilde{B}_1$, we have

$$\begin{aligned}
e^T(k)\Theta e(k) &\leq \varepsilon + \varrho(\bar{x}(k) - \tilde{x}(k))^T \Theta(\bar{x}(k) - \tilde{x}(k)) \\
&= \varepsilon + \varrho \begin{bmatrix} \tilde{x}(k) \\ \bar{x}(k) \end{bmatrix}^T \begin{bmatrix} \tilde{B}_1^T P \tilde{B}_1 & -\tilde{B}_1^T P \tilde{B}_1 \\ -\tilde{B}_1^T P \tilde{B}_1 & \tilde{B}_1^T P \tilde{B}_1 \end{bmatrix} \begin{bmatrix} \tilde{x}(k) \\ \bar{x}(k) \end{bmatrix} \\
&= \varepsilon + \varrho \begin{bmatrix} \xi(k) \\ \eta(k) \end{bmatrix}^T \begin{bmatrix} \Lambda_{11} & \Lambda_{12} \\ \Lambda_{12}^T & \Lambda_{22} \end{bmatrix} \begin{bmatrix} \xi(k) \\ \eta(k) \end{bmatrix} \\
&\leq \varepsilon + \xi^T(k)(\Lambda_{11} + I_{2(n+p)})\xi(k) + \eta^T(k)(\Lambda_{12}^T \Lambda_{12} + \Lambda_{22})\eta(k),
\end{aligned}$$

where $\eta(k) = [d^T(k-1), \ d^T(k)]^T$, and $\Lambda_{12} = \begin{bmatrix} -\tilde{B}_1^T P \tilde{B}_{12} \\ \tilde{B}_{11}^T P \tilde{B}_{12} \end{bmatrix}$, $\Lambda_{22} = \tilde{B}_{12}^T P \tilde{B}_{12}$.

It follows from $\|d(k)\| \leq \bar{d}$ and $\|h(k)\| \leq \delta_d$ that $\|f(k)\|^2 = \|h(k)\|^2 + \|d(k)\|^2 \leq \delta_d^2 + \bar{d}^2$. Then

$$\begin{aligned}
\Delta V(k) &\leq \xi^T(k)(3\tilde{A}^T P\tilde{A} - P + 3(\Lambda_{11} + I_{2(n+p)}))\xi(k) + 3\varepsilon \\
&\quad + \eta^T(k)(\Lambda_{12}^T \Lambda_{12} + \Lambda_{22})\eta(k) + 3f^T(k)\tilde{B}_2^T P\tilde{B}_2 f(k) \\
&\leq \xi^T(k)(3\tilde{A}^T P\tilde{A} - P + 3(\Lambda_{11} + I_{2(n+p)}))\xi(k) + 3\varepsilon \\
&\quad + 2\|\Lambda_{12}^T \Lambda_{12} + \Lambda_{22}\|\bar{d}^2 + 3\|\tilde{B}_2^T P\tilde{B}_2\|(\delta_d^2 + \bar{d}^2).
\end{aligned}$$

Let $\mu = 3\varepsilon + 2\|\Lambda_{12}^T \Lambda_{12} + \Lambda_{22}\|\bar{d}^2 + 3\|\tilde{B}_2^T P\tilde{B}_2\|(\delta_d^2 + \bar{d}^2)$. For $k \in [k_i, k_{i+1})$, $\Delta V(k)$ is obtained as

$$\begin{aligned}
\Delta V(k) &\leq -\lambda_{\min}(-Q)\|\xi(k)\|^2 + \mu, \\
&\leq -\frac{\lambda_{\min}(-Q)}{\lambda_{\max}(P)}\xi^T(k)P\xi(k) + \mu, \\
&= -\delta V(k) + \mu. \tag{12.14}
\end{aligned}$$

According to the result of (12.14), $V(k)$ decays in $[k_i, k_{i+1})$, $i = 1, 2, \ldots$. It follows from (12.11) that $\xi(k_{i+1}) = \tilde{A}\xi(k_{i+1} - 1) + \tilde{B}_1 e(k_{i+1} - 1) + \tilde{B}_2 f(k_{i+1} - 1)$, and similarly, we have $V(k_{i+1}) - V(k_{i+1} - 1) \leq -\delta V(k_{i+1} - 1) + \mu$, which gives $V(k_{i+1}) \leq (1 - \delta)V(k_{i+1} - 1) + \mu$. This together with (12.14) implies that

$$V(k_{i+1}) \leq (1 - \delta)^{k_{i+1} - k_i}V(k_i) + \sum_{j=0}^{k_{i+1} - k_i - 1}(1 - \delta)^j \mu.$$

Thus, $V(k) \leq (1 - \delta)^{k - k_0}V(k_0) + \frac{\mu}{\delta}$, and

$$\|\xi(k)\|^2 \leq (1 - \delta)^{k - k_0}\frac{\lambda_{\max}(P)}{\lambda_{\min}(P)}\|\xi(k_0)\|^2 + \frac{\mu}{\delta\lambda_{\min}(P)}. \tag{12.15}$$

Considering the fact that $\delta = \frac{\lambda_{\min}(-Q)}{\lambda_{\max}(P)} < 1$, we have $\lim_{k \to \infty}(1 - \delta)^{(k - k_0)} = 0$. Thus, with the updating instants (12.6), the states of system (12.11) are UUB. □

Remark 12.6. *For the two user-defined parameters ε and ϱ in condition (12.6), (1) the ultimate upper bound of the states of closed-loop system will be affected by parameter ε and (2) the feasibility of condition (12.13) and thus the design parameter Θ will be affected by parameter ϱ. Therefore, by suitably choosing the parameters ε and ϱ, we can obtain a better system performance and save more communication resources.*

Remark 12.7. *Note that matrix $\tilde{A}$ can be Schur by choosing suitable L and K in (12.13). Then, there always exists matrix P such that condition (12.13) holds for given small parameter ϱ.*

In order to achieve the active disturbance rejection, the disturbance compensation gain K_d will be designed here.

Theorem 12.2. *For the given controller gain K_x such that $A + BK_x$ is Schur, there always exists a disturbance rejection gain K_d if equation*

$$\text{rank} \left[C(I_n - (A + BK_x))^{-1}B \right]$$
$$= \text{rank} \left[C(I_n - (A + BK_x))^{-1}B, -C(I_n - (A + BK_x))^{-1}B_d \right] \quad (12.16)$$

holds. In addition, if the rank condition (12.16) is true, then the disturbance rejection gain can be calculated as

$$K_d = - \left(C(I_n - (A + BK_x))^{-1}B \right)^{-1} C(I_n - (A + BK_x))^{-1}B_d. \quad (12.17)$$

Proof. Noting that $h(k) = d(k + 1) - 2d(k) + d(k - 1)$ and matrix $\bar{A} - L\bar{C}$ is Schur, under Assumption 12.3, we can conclude that $\tilde{\bar{x}}(k) \in O(T_s^3)$ since $h(k) \in O(T_s^3)$ at the steady-state. Thus, according to (12.9) and (12.10), we have the following equation at the steady-state:

$$x(k + 1) = (A + BK_x)x(k) - BK\tilde{\bar{x}}(k) - BKe(k) + (B_d + BK_d)d(k). \quad (12.18)$$

Taking Z-transform to system (12.18), we get the transfer function from $d(k)$ to $y(k)$ as $G_{yd}(z) = C(zI_n - (A + BK_x))^{-1}(BK_d + B_d)$.

To reject the disturbance in the output channels at the steady-state, we can set $C(I_n - (A + BK_x))^{-1}(BK_d + B_d) = 0$, which gives

$$C(I_n - (A + BK_x))^{-1}BK_d = -C(I_n - (A + BK_x))^{-1}B_d. \quad (12.19)$$

So the disturbance compensation gain K_d is solved by equation (12.19) when the rank condition (12.16) holds. Furthermore, it follows from Assumption 12.2 that $C(I_n - (A + BK_x))^{-1}B$ is invertible. Hence, K_d can be computed from (12.19) as the one given in (12.17). □

Note that matrices $A + BK_x$ and $\bar{A} - L\bar{C}$ can be guaranteed to be Schur by choosing suitable feedback control gain K_x and observer gain L, respectively. Then, $C(I - (A + BK_x))^{-1}B$ is invertible from Assumption 12.2. Therefore, it turns out that disturbance $d(k)$ can be compensated by the proposed controller (12.5) in the output channels at the steady-state.

Theorem 12.3. *In the output channel of system (12.1), the bounded disturbance $d(k)$ can be removed at the steady-state when controller (12.5) is employed in system (12.1).*

Proof. Taking Z-transform to system (12.18) yields

$$X(z) = (zI_n - (A + BK_x))^{-1}zx(0) - (zI_n - (A + BK_x))^{-1}BKE(z)$$
$$+ (zI_n - (A + BK_x))^{-1}(BK_d + B_d)D(z). \quad (12.20)$$

It follows readily from (12.1) and (12.20) that the output can be represented as

$$
\begin{aligned}
Y(z) &= CX(z) \\
&= C(zI_n - (A + BK_x))^{-1}zx(0) - C(zI_n - (A + BK_x))^{-1}BKE(z) \\
&\quad + C(zI_n - (A + BK_x))^{-1}(BK_d + B_d)D(z).
\end{aligned}
\tag{12.21}
$$

It is easily seen from (12.21) that with the choice of K_d, the disturbance $d(k)$ is removed from $y(k)$ in the steady-state. □

Remark 12.8. *In order to ensure $A + BK_x$ to be Schur, the controller gain K_x can be chosen with any existing design methods, e.g., the pole placement technique.*

Remark 12.9. *The disturbance compensation gain K_d can be chosen as $K_d = -1$ if $B = B_d$. While, K_d can be designed as $K_d = -[C(I_n - (A + BK_x))^{-1}B]^{-1}C(I_n - (A + BK_x))^{-1}B_d$ if the disturbance is the mismatched one.*

12.4 REDUCED-ORDER ESO-BASED EVENT-TRIGGERED DISTURBANCE REJECTION CONTROL

In the previous section, the full-order ESO case has been considered. Clearly, this is useful if the knowledge of states $x(k)$ and the external disturbance $d(k)$ are unavailable. However, in practice, some states are measured from the outputs directly and thus it is not necessary to estimate the available states. In this case, it is advisable to design the observer with reduced order.

Noting that rank$(C) = q$, we assume $C = [I_q, \ 0]$ for simplicity, and partition A, B, B_d as

$$
A = \begin{bmatrix} A_{11} & A_{12} \\ A_{21} & A_{22} \end{bmatrix}, \quad B = \begin{bmatrix} B_1 \\ B_2 \end{bmatrix}, \quad B_d = \begin{bmatrix} B_{1d} \\ B_{2d} \end{bmatrix},
$$

respectively, where $A_{11} \in \mathbb{R}^{q \times q}$, $A_{12} \in \mathbb{R}^{q \times (n-q)}$, $A_{21} \in \mathbb{R}^{(n-q) \times q}$, $A_{22} \in \mathbb{R}^{(n-q) \times (n-q)}$, $B_{1d} \in \mathbb{R}^{q \times p}$, $B_{2d} \in \mathbb{R}^{(n-q) \times p}$, $B_1 \in \mathbb{R}^{q \times m}$, and $B_2 \in \mathbb{R}^{(n-q) \times m}$. Thus, we have the following partition of $\bar{A}$, $\bar{A}_d$, and $\bar{B}$,

$$
\bar{A} = \begin{bmatrix} A_{11} & A_{12} & 0 & B_{1d} \\ A_{21} & A_{22} & 0 & B_{2d} \\ 0 & 0 & 0 & I_p \\ 0 & 0 & -I_p & 2I_p \end{bmatrix} \triangleq \begin{bmatrix} \bar{A}_{11} & \bar{A}_{12} \\ \bar{A}_{21} & \bar{A}_{22} \end{bmatrix}, \quad \bar{B} = \begin{bmatrix} B_1 \\ B_2 \\ 0 \\ 0 \end{bmatrix} \triangleq \begin{bmatrix} \bar{B}_1 \\ \bar{B}_2 \end{bmatrix}.
$$

Then, we take the state transformation $\bar{\omega}(k) = T\bar{x}(k)$ for system (12.3), where $T = \begin{bmatrix} I_q & 0 \\ L_r & I_{n+2p-q} \end{bmatrix}$ and let $\bar{\omega}(k) = [\bar{\omega}_1^T(k), \bar{\omega}_2^T(k)]^T$ with $\bar{\omega}_1(k) \in \mathbb{R}^q$, $\bar{\omega}_2(k) \in \mathbb{R}^{n+2p-q}$, we can get

$$
\begin{aligned}
\bar{\omega}(k+1) &= \hat{A}\bar{\omega}(k) + \hat{B}u(k) + \hat{E}h(k), \\
y(k) &= \bar{\omega}_1(k),
\end{aligned}
\tag{12.22}
$$

where

$$\hat{A} = T\bar{A}T^{-1} = \left[\begin{array}{c|ccc} \Upsilon_{11} & A_{12} & 0 & B_{1d} \\ \hline \Upsilon_{21} & L_{1r}A_{12} + A_{22} & 0 & L_{1r}B_{1d} + B_{2d} \\ \Upsilon_{31} & L_{2r}A_{12} & 0 & L_{2r}B_{1d} + I_p \\ \Upsilon_{41} & L_{3r}A_{12} & -I_p & L_{3r}B_{1d} + 2I_p \end{array} \right] \triangleq \left[\begin{array}{c|c} \hat{A}_{11} & \hat{A}_{12} \\ \hline \hat{A}_{21} & \hat{A}_{22} \end{array} \right],$$

$$\hat{B} = T\bar{B} = \left[\begin{array}{c} B_1 \\ \hline L_{1r}B_1 + B_2 \\ L_{2r}B_1 \\ L_{3r}B_1 \end{array} \right] \triangleq \left[\begin{array}{c} \hat{B}_1 \\ \hline \hat{B}_2 \end{array} \right], \hat{E} = TE = \left[\begin{array}{c} 0 \\ 0 \\ 0 \\ I_p \end{array} \right] \triangleq \left[\begin{array}{c} 0 \\ \hline \hat{E}_2 \end{array} \right],$$

with $\Upsilon_{11} = A_{11} - A_{12}L_{1r} - B_{1d}L_{3r}$, $\Upsilon_{21} = L_{1r}A_{11} + A_{21} - L_{1r}A_{12}L_{1r} - A_{22}L_{1r} - L_{1r}B_{1d}L_{3r} - B_{2d}L_{3r}$, $\Upsilon_{31} = L_{2r}A_{11} - L_{2r}A_{12}L_{1r} - L_{2r}B_{1d}L_{3r} - L_{3r}$, and $\Upsilon_{41} = L_{3r}A_{11} - L_{3r}A_{12}L_{1r} + L_{2r} - L_{3r}B_{1d}L_{3r} - 2L_{3r}$. $L_r = \left[\begin{array}{c} L_{1r} \\ L_{2r} \\ L_{3r} \end{array} \right] \in \mathbb{R}^{(n+2p-q)\times q}$ is

selected such that $\bar{A}_{22} + L_r\bar{A}_{12}$ is Schur and $L_{1r} \in \mathbb{R}^{(n-q)\times q}$, $L_{2r} \in \mathbb{R}^{p\times q}$, $L_{3r} \in \mathbb{R}^{p\times q}$.

According to system (12.22), it is obtained that $\bar{\omega}_2(k+1) = \hat{A}_{22}\bar{\omega}_2(k) + \hat{B}_2u(k) + \hat{A}_{21}y(k) + \hat{E}_2h(k)$. Then, a reduced-order ESO is designed as

$$\hat{\bar{\omega}}_2(k+1) = \hat{A}_{22}\hat{\bar{\omega}}_2(k) + \hat{B}_2u(k) + \hat{A}_{21}y(k), \tag{12.23}$$

where $\hat{\bar{\omega}}_2(k)$ is the estimate of reduced-order state $\bar{\omega}_2(k)$.

Lemma 12.2. *Under Assumption 12.2, pair $(\bar{A}_{22}, \bar{A}_{12})$ is detectable.*

Proof. From the proof of Lemma 12.1, it is known that if Assumption 12.2 is satisfied, then pair $(\bar{A}, \bar{C})$ is detectable. Thus, we have that, $\forall z \in \mathbb{C}, |z| \geq 1$,

$$\text{rank} \left[\begin{array}{c} zI_{n+2p} - \bar{A} \\ \bar{C} \end{array} \right] = \text{rank} \left[\begin{array}{cc} zI_q - \bar{A}_{11} & -\bar{A}_{12} \\ -\bar{A}_{21} & zI_{n+2p-q} - \bar{A}_{22} \\ I_q & 0 \end{array} \right]$$

$$= n + 2p,$$

which means that $\left[\begin{array}{c} -\bar{A}_{12} \\ zI_{n+2p-q} - \bar{A}_{22} \end{array} \right]$ is also of full column rank, i.e.,

$$\text{rank} \left[\begin{array}{c} \bar{A}_{12} \\ zI_{n+2p-q} - \bar{A}_{22} \end{array} \right] = n + 2p - q, \forall z \in \mathbb{C}, |z| \geq 1.$$

Hence, we can conclude that pair $(\bar{A}_{22}, \bar{A}_{12})$ is detectable, and $\bar{A}_{22} + L_r\bar{A}_{12}$ can be Schur by choosing a gain matrix L_r. This completes the proof. □

Remark 12.10. *Note that the states which can be obtained from system outputs directly will not be estimated by the reduced-order ESO. Therefore, the dimensions of all matrices in (12.22) are lower than the ones in (12.4), which leads to a lower computational complexity in practical cases.*

Noticing that T can be repartitioned into $T = \begin{bmatrix} T_1 & 0 \\ T_3 & I_{2p} \end{bmatrix}$ with $T_1 \in \mathbb{R}^{n \times n}$ and $T_3 \in \mathbb{R}^{2p \times n}$, and it is easy to obtain that

$$T_1 = \begin{bmatrix} I_q & 0 \\ L_{1r} & I_{n-q} \end{bmatrix}, T_3 = \begin{bmatrix} L_{2r} & 0 \\ L_{3r} & 0 \end{bmatrix}.$$

Furthermore, we can also partition $\bar{\omega}(k)$ as $\bar{\omega}(k) = [\omega^T(k), w^T(k)]^T$ with $\omega(k) \in \mathbb{R}^n$, $w(k) \in \mathbb{R}^{2p}$, and thus $\omega(k) = T_1 x(k)$, $w(k) = T_3 x(k) + \eta(k)$, which implies that $w(k) = T_3 T_1^{-1} \omega(k) + \eta(k)$ with $\eta(k) = [d^T(k-1), d^T(k)]^T$. Accordingly, $\hat{A}$, $\hat{B}$ and $\hat{E}$ can be repartitioned into

$$\hat{A} = \left[\begin{array}{cc|cc} \Upsilon_{11} & A_{12} & 0 & B_{1d} \\ \Upsilon_{21} & L_{1r}A_{12} + A_{22} & 0 & L_{1r}B_{1d} + B_{2d} \\ \hline \Upsilon_{31} & L_{2r}A_{22} & 0 & L_{2r}B_{1d} + I_p \\ \Upsilon_{41} & L_{3r}A_{12} & -I_p & L_{3r}B_{1d} + 2I_p \end{array} \right] \triangleq \left[\begin{array}{c|c} \tilde{A}_{11} & \tilde{A}_{12} \\ \hline \tilde{A}_{21} & \tilde{A}_{22} \end{array} \right],$$

$$\hat{B} = \left[\begin{array}{c} B_1 \\ L_{1r}B_1 + B_2 \\ \hline L_{2r}B_1 \\ L_{3r}B_1 \end{array} \right] \triangleq \left[\begin{array}{c} \tilde{B}_1 \\ \hline \tilde{B}_2 \end{array} \right], \hat{E} = \left[\begin{array}{c} 0 \\ 0 \\ \hline 0 \\ I_p \end{array} \right] \triangleq \left[\begin{array}{c} 0 \\ \hline \tilde{E}_2 \end{array} \right].$$

For simplicity, we further partition $w(k)$ as $w(k) = \begin{bmatrix} w_1^T(k) & w_2^T(k) \end{bmatrix}^T$ with $w_1(k) \in \mathbb{R}^p$ and $w_2(k) \in \mathbb{R}^p$ and define $\tilde{B}_d \triangleq \begin{bmatrix} B_{1d} \\ L_{1r}B_{1d} + B_{2d} \end{bmatrix}$. It is easy to obtain that $w_1(k) = L_{2r}y(k) + d(k-1)$ and $w_2(k) = L_{3r}y(k) + d(k)$.

Thus, we can arrive at

$$\begin{aligned} \omega(k+1) &= \tilde{A}_{11}\omega(k) + \tilde{A}_{12}w(k) + \tilde{B}_1 u(k) \\ &= \tilde{A}_{11}\omega(k) + \tilde{B}_d w_2(k) + \tilde{B}_1 u(k) \\ &= (\tilde{A}_{11} + \tilde{B}_d L_{3r}C)\omega(k) + \tilde{B}_1 u(k) + \tilde{B}_d d(k). \end{aligned} \tag{12.24}$$

Similarly, to reduce the updating frequency of control signals, estimated states from the reduced-order observer will be transmitted to the controller at each updating instant. The control input is held until the next signal updating by applying a ZOH. Thus, similar to (12.5), the reduced-order observer-based event-triggered controller is written as follows:

$$u(k) = K_l \hat{\bar{\omega}}(k_i) = K_r \hat{\omega}(k_i) + K_{rd}\hat{w}_2(k_i), \ k \in [k_i, k_{i+1}), \tag{12.25}$$

where vector $\hat{\bar{\omega}}(k) = [\hat{\omega}^T(k) \ \hat{w}^T(k)]^T$ denotes the estimate of $\bar{\omega}(k)$, $\hat{\omega}(k) \in \mathbb{R}^n$ is the estimate of $\omega(k)$, and $\hat{w}(k) = [\hat{w}_1^T(k) \ \hat{w}_2^T(k)]^T \in \mathbb{R}^{2p}$ is the estimate of $w(k)$, and $K_l = \begin{bmatrix} K_r & 0 & K_{rd} \end{bmatrix}$ is the controller gain with appropriate dimension. The updating instants k_i, $i = 0, 1, 2, ...$, of controller (12.25) are determined by

$$k_{i+1} = \sup \left\{ k > k_i \left| \hat{e}^T(k)\Theta_r\hat{e}(k) \le \varepsilon + \varrho\hat{\bar{\omega}}^T(k)\Theta_r\hat{\bar{\omega}}(k) \right. \right\}, k \in \mathbb{N}, \tag{12.26}$$

where the error signal is $\hat{e}(k) = [\hat{e}_1^T(k), \hat{e}_2^T(k)]$ with $\hat{e}_1(k) = \hat{\omega}(k) - \hat{\omega}(k_i)$, $\hat{e}_2(k) = \hat{w}(k) - \hat{w}(k_i)$, ε and ϱ are user-defined parameters, and Θ_r is a designed weighting matrix. Note that $\hat{w}_2(k) = L_{3r}y(k) + \hat{d}(k)$ and $y(k) = C\omega(k)$. Thus, $y(k) - y(k_i) = C\hat{e}_1(k)$. It can be seen that, at updating instants, $\hat{e}(k_i) \equiv 0$. Under the definition of updating instants k_i in (12.26), in the inter-event interval $[k_i, k_{i+1})$, the inequality $\hat{e}^T(k)\Theta_r\hat{e}(k) \leq \varepsilon + \varrho\hat{\bar{\omega}}^T(k)\Theta_r\hat{\bar{\omega}}(k)$ always holds.

Next, we will show that in the output channel, the disturbance can be rejected at the steady-state with a properly designed disturbance compensation gain.

It follows from (12.22) that $\bar{\omega}_1(k) = \hat{\bar{\omega}}_1(k) = y(k)$. Let the estimation error be $\tilde{\bar{\omega}}(k) = \bar{\omega}(k) - \hat{\bar{\omega}}(k)$, and the event-triggered controller (12.25) can be rewritten as

$$u(k) = (K_r + K_{rd}L_{3r}C)\hat{\omega}(k_i) + K_{rd}\hat{d}(k_i), \ k \in [k_i, k_{i+1}). \tag{12.27}$$

Substituting (12.27) into system (12.24) gives that

$$\begin{aligned}
\omega(k+1) &= (\tilde{A}_{11} + \tilde{B}_d L_{3r}C)\omega(k) + \tilde{B}_1 u(k) + \tilde{B}_d d(k)\\
&= (\tilde{A}_{11} + \tilde{B}_d L_{3r}C + \tilde{B}_1(K_r + K_{rd}L_{3r}C))\omega(k)\\
&\quad - \tilde{B}_1(K_r + K_{rd}L_{3r}C))(\hat{e}_1(k) + \tilde{\omega}(k)) + (\tilde{B}_d + \tilde{B}_1 K_{rd})d(k)\\
&\quad - \tilde{B}_1 K_{rd}(\hat{d}(k) - \hat{d}(k_i) + d(k))\\
&= (\tilde{A}_{11} + \tilde{B}_1 K_r + (\tilde{B}_d + \tilde{B}_1 K_{rd})L_{3r}C)\omega(k)\\
&\quad - \tilde{B}_1 K_l\hat{e}(k) - \tilde{B}_1 K_l \begin{bmatrix} 0 \\ I_{n+2p-q} \end{bmatrix} \tilde{\bar{\omega}}_2(k) + (\tilde{B}_d + \tilde{B}_1 K_{rd})d(k),
\end{aligned}$$

where the last equality uses the fact that $\tilde{\bar{\omega}}(k) = \begin{bmatrix} 0 \\ I_{n+2p-q} \end{bmatrix} \tilde{\bar{\omega}}_2(k)$.

It is easy to obtain that $\tilde{\bar{\omega}}_2(k+1) = \bar{\omega}_2(k+1) - \hat{\bar{\omega}}_2(k+1) = \hat{A}_{22}\tilde{\bar{\omega}}_2(k) + \hat{E}_2 h(k)$. Define the augmented variable as $\xi_r(k) = [\tilde{\bar{\omega}}_2^T(k), \omega^T(k)]^T$, and the closed-loop system is obtained as

$$\xi_r(k+1) = \tilde{A}_r \xi_r(k) + \tilde{B}_{1r}\hat{e}(k) + \tilde{B}_{2r}g(k), \ k \in [k_i, k_{i+1}), \tag{12.28}$$

where

$$\tilde{B}_{1r} = \begin{bmatrix} 0 \\ -\tilde{B}_1 K_l \end{bmatrix}, \ g(k) = \begin{bmatrix} h(k) \\ d(k) \end{bmatrix}, \ \tilde{B}_{2r} = \begin{bmatrix} \tilde{E}_2 & 0 \\ 0 & \tilde{B}_d + \tilde{B}_1 K_{rd} \end{bmatrix},$$

$$\tilde{A}_r = \begin{bmatrix} \hat{A}_{22} & 0 \\ -\tilde{B}_1 K_l \begin{bmatrix} 0 \\ I_{n+2p-q} \end{bmatrix} & \tilde{A}_{11r} \end{bmatrix},$$

with $\tilde{A}_{11r} = \tilde{A}_{11} + \tilde{B}_1 K_r + (\tilde{B}_d + \tilde{B}_1 K_{rd})L_{3r}C$.

Theorem 12.4. *With the UIS (12.26), the states of system (12.28) are UUB if there exists $P > 0$ satisfying*

$$3\tilde{A}_r^T P \tilde{A}_r - P + 3\varrho(\tilde{\Lambda}_{22r} + I_{2(n+p)-q}) < 0, \tag{12.29}$$

where

$$\tilde{\Lambda}_{22r} = \begin{bmatrix} \tilde{B}_{12qr}^T P \tilde{B}_{12qr} & -\tilde{B}_{12qr}^T P \tilde{B}_{11r} \\ -\tilde{B}_{11r}^T P \tilde{B}_{12qr} & \tilde{B}_{11r}^T P \tilde{B}_{11r} \end{bmatrix}, \tilde{B}_{11r} = \begin{bmatrix} 0 \\ -\tilde{B}_1 K_r \end{bmatrix},$$

$$\tilde{B}_{12r} = \begin{bmatrix} 0 & 0 \\ 0 & -\tilde{B}_1 K_{rd} \end{bmatrix}, \tilde{B}_{11qr} = \begin{bmatrix} 0 \\ -\tilde{B}_1 K_{qr} \end{bmatrix}, \tilde{B}_{12qr} = \begin{bmatrix} 0 \\ -\tilde{B}_1 K_{qrd} \end{bmatrix},$$

with $\tilde{B}_{1r} = [\tilde{B}_{11r}\ \tilde{B}_{12r}] = [\tilde{B}_{11qr}\ \tilde{B}_{12qr}]$, $K_l = [K_{qr}\ K_{qrd}]$, $\tilde{B}_{11r} \in \mathbb{R}^{(2n+2p-q)\times q}$, $\tilde{B}_{12r} \in \mathbb{R}^{(2n+2p-q)\times(n+2p-q)}$, $\tilde{B}_{11qr} \in \mathbb{R}^{(2n+2p-q)\times n}$, $\tilde{B}_{12qr} \in \mathbb{R}^{(2n+2p-q)\times 2p}$, $K_{qr} \in \mathbb{R}^{m\times q}$, and $K_{qrd} \in \mathbb{R}^{m\times(n+2p-q)}$, and the weighting matrix in (12.26) is chosen as $\Theta_r = \tilde{B}_{1r}^T P \tilde{B}_{1r}$.

Proof. The proof is similar to the proof of Theorem 12.1 and the details are omitted here for brevity. □

Remark 12.11. *The gain matrices K_r and L_r should be chosen such that matrices $\tilde{A}_{11} + (\tilde{B}_d + \tilde{B}_1 K_{rd})L_{3r}C + \tilde{B}_1 K_r$ and $\bar{A}_{22} + L_r \bar{A}_{12}$ are Schur, respectively.*

Theorem 12.5. *If the rank condition*

$$rank[B] = rank[B,\ B_d]. \tag{12.30}$$

holds, then there always exists K_{rd} designed by

$$K_{rd} = -(\tilde{B}_1^T \tilde{B}_1)^{-1} \tilde{B}_1^T \tilde{B}_d. \tag{12.31}$$

With the choice of K_{rd}, the bounded disturbance is rejected from the output channel of system (12.22) under control law (12.25).

Proof. Similarly, it can be shown that $\tilde{\bar{\omega}}_2(k) \in O(T_s^3)$ at the steady-state, so we have the following equation at the steady-state,

$$\omega(k+1) = (\tilde{A}_{11} + \tilde{B}_1 K_r + (\tilde{B}_d + \tilde{B}_1 K_{rd})L_{3r}C)\omega(k)$$
$$- \tilde{B}_1 K_l \hat{e}(k) + (\tilde{B}_d + \tilde{B}_1 K_{rd})d(k). \tag{12.32}$$

Then, we obtain that

$$\Omega(z) = (zI_n - (\tilde{A}_{11} + \tilde{B}_1 K_r + (\tilde{B}_d + \tilde{B}_1 K_{rd})L_{3r}C))^{-1} z\omega(0)$$
$$+ (zI_n - (\tilde{A}_{11} + \tilde{B}_1 K_r + (\tilde{B}_d + \tilde{B}_1 K_{rd})L_{3r}C))^{-1}(\tilde{B}_d + \tilde{B}_1 K_{rd})D(z)$$
$$- (zI_n - (\tilde{A}_{11} + \tilde{B}_1 K_r + (\tilde{B}_d + \tilde{B}_1 K_{rd})L_{3r}C))^{-1} \tilde{B}_1 K_l \hat{E}(z). \tag{12.33}$$

Based on (12.22), (12.33), and the choice of K_{rd}, the output can be represented as

$$Y(z) = \bar{\Omega}_1(k) = C\Omega(z)$$
$$= C(zI_n - (\tilde{A}_{11} + \tilde{B}_1 K_r + (\tilde{B}_d + \tilde{B}_1 K_{rd})L_{3r}C))^{-1}(\tilde{B}_d + \tilde{B}_1 K_{rd})D(z)$$
$$- C(zI_n - (\tilde{A}_{11} + \tilde{B}_1 K_r + (\tilde{B}_d + \tilde{B}_1 K_{rd})L_{3r}C))^{-1} \tilde{B}_1 K_l \hat{E}(z). \tag{12.34}$$

From (12.34), it is shown that to reject the disturbance from the output channel, $\tilde{B}_d + \tilde{B}_1 K_{rd} = 0$ is provided, which implies that

$$\tilde{B}_1 K_{rd} = -\tilde{B}_d. \tag{12.35}$$

Under the condition

$$\text{rank}\left[\tilde{B}_1\right] = \text{rank}\left[\tilde{B}_1, \ -\tilde{B}_d\right], \tag{12.36}$$

it is known that matrix equation (12.35) admits a solution K_{rd} given by (12.31). It can be shown that condition (12.36) is equivalent to condition (12.30) by noticing the equations $\tilde{B}_d = \mathcal{L}_1 B_d$, $\tilde{B}_1 = \mathcal{L}_1 B$, where $\mathcal{L}_1 = \begin{bmatrix} I_q & 0 \\ L_{1r} & I_{n-q} \end{bmatrix}$, which completes the proof. $\qquad\square$

12.5 REDUCED-ORDER EFSO-BASED EVENT-TRIGGERED DISTURBANCE REJECTION CONTROL

In the previous section, the reduced-order ESO case has been considered. It is clear that some system transformations are necessary for designing the reduced-order ESO. This section aims to introduce the reduced-order extended functional state observer (EFSO) to simplify the design procedures of the observer.

In this section, we denote $x(k) \triangleq \begin{bmatrix} y(k) \\ H_0 x(k) \end{bmatrix}$, where $H_0 \in \mathbb{R}^{(n-q)\times n}$ satisfies $\text{rank}(H_0) = n - q$ and

$$\begin{bmatrix} C \\ H_0 \end{bmatrix} = I_n. \tag{12.37}$$

It is clear that $H_0 x(k)$ is the unavailable state which cannot be measured from $y(k)$ directly. To achieve the active disturbance rejection, the following reduced-order EFSO is proposed to obtain the unavailable states and the disturbance estimates,

$$\begin{aligned} \varpi(k+1) &= L_1\varpi(k) + L_2 y(k) + L_3 u(k) \\ \hat{\nu}(k) &= L_4\varpi(k) + L_5 y(k), \end{aligned} \tag{12.38}$$

where $\varpi(k) \in \mathbb{R}^{n+2p-q}$ is the intermediate variable, $\hat{\nu}(k) = [(H_0\hat{x}(k))^T, \ \hat{d}^T(k-1), \ \hat{d}^T(k)]^T$ with $H_0\hat{x}(k)$ and $\hat{d}(k)$ are the estimates of unavailable states and disturbance, respectively.

For simplicity, matrix $\bar{A}$ is partitioned as $\bar{A} \triangleq \begin{bmatrix} \bar{A}_{11} & \bar{A}_{12} \\ \bar{A}_{21} & \bar{A}_{22} \end{bmatrix}$, $\bar{A}_{11} \in \mathbb{R}^{q\times q}$, $\bar{A}_{12} \in \mathbb{R}^{q\times(n+2p-q)}$, $\bar{A}_{21} \in \mathbb{R}^{(n+2p-q)\times q}$, $\bar{A}_{22} \in \mathbb{R}^{(n+2p-q)\times(n+2p-q)}$. The observer parameters L_1, L_2, and L_3 are designed as

$$L_1 = \bar{A}_{22} - L_5\bar{A}_{12}, \tag{12.39}$$

$$L_2 = L_1 L_4^{-1} L_5 - L_4^{-1} L_5\bar{A}_{11} + L_4^{-1}\bar{A}_{21}, \tag{12.40}$$

$$L_3 = L_4^{-1}(H - L_5\bar{C})\bar{B}, \tag{12.41}$$

and $L_4 = \lambda I_{n+2p-q}$ with that $\lambda \neq 0$ is a designed parameter, and L_5 can be chosen such that L_1 is Schur by the pole assignment method, and

$$H = \begin{bmatrix} H_0 & 0 & 0 \\ 0 & I_p & 0 \\ 0 & 0 & I_p \end{bmatrix}. \tag{12.42}$$

Defining the estimation error as $\epsilon(k) = H\bar{x}(k) - \hat{\nu}(k)$, we can obtain that

$$\begin{aligned}
\epsilon(k+1) &= H\bar{x}(k+1) - \hat{\nu}(k+1) \\
&= L_1\epsilon(k) + (\bar{H}\bar{A} - L_1\bar{H} - L_4L_2\bar{C})\bar{x}(k) \\
&\quad + (\bar{H}\bar{B} - L_4L_3)u(k) + \bar{H}Eh(k),
\end{aligned} \tag{12.43}$$

where $\bar{H} = H - L_5\bar{C}$. Taking (12.39), (12.40), and (12.41) into (12.43) gives that

$$\begin{aligned}
\bar{H}\bar{A} - L_1\bar{H} - L_4L_2\bar{C} &= (H - L_5\bar{C})\bar{A} - L_1(H - L_5\bar{C}) - L_4L_2\bar{C} \\
&= H\bar{A} - L_5\bar{C}\bar{A} - L_1H + L_5\bar{A}_{11}\bar{C} - \bar{A}_{21}\bar{C} \\
&= \begin{bmatrix} -L_5 & I_{n+2p-q} \end{bmatrix} \left(\begin{bmatrix} \bar{C} \\ H \end{bmatrix} \bar{A} - \begin{bmatrix} \bar{A}_{11} \\ \bar{A}_{21} \end{bmatrix} \bar{C} \right) - L_1H \\
&= \begin{bmatrix} -L_5 & I_{n+2p-q} \end{bmatrix} \begin{bmatrix} \bar{A}_{12} \\ \bar{A}_{22} \end{bmatrix} H - L_1H \\
&= -(L_1 - \bar{A}_{22} + L_5\bar{A}_{12})H \\
&= 0,
\end{aligned}$$

and $\bar{H}\bar{B} - L_4L_3 = (H - L_5\bar{C})\bar{B} - L_4L_3 = 0$. Thus, (12.43) can be rewritten as

$$\epsilon(k+1) = L_1\epsilon(k) + \bar{H}Eh(k). \tag{12.44}$$

According to (12.37), it can be verified that

$$\hat{x}(k) = \begin{bmatrix} y(k) \\ [I_{n+2p-q}, \; 0]\,\hat{\nu}(k) \end{bmatrix}, \tag{12.45}$$

$$\hat{d}(k) = \begin{bmatrix} 0 & I_p \end{bmatrix} \hat{\nu}(k). \tag{12.46}$$

Thus, the event-triggered disturbance rejection controller is designed as (12.5). For simplicity of the following discussions, we define $\mathcal{K}_x \triangleq [K_c, \; K_h]$, $K_\nu \triangleq [K_h, \; 0, \; K_d]$, and $\mathcal{K} \triangleq [K_c, \; K_\nu]$, where $K_c \in \mathbb{R}^{m \times q}$, and $K_h \in \mathbb{R}^{m \times (n+2p-q)}$, and it can be verified that controller (12.5) can also be rewritten as

$$u(k) = K_c y(k_i) + K_\nu \hat{\nu}(k_i), \; k \in [k_i, \; k_{i+1}). \tag{12.47}$$

Remark 12.12. *Referring to [205], [208], and [140], the disturbance rejection control methods are presented to counteract the external disturbance and improve the control performance in CPSs. To this end, a RESFO (12.38) is designed to estimate the* *unknown states and disturbance of the physical plants. Note that the observer (12.38) is in a reduced-order form and simpler to be designed comparing with [139], and the controller (12.47) is designed by using the outputs of both system (12.1) and the estimates obtained by observer (12.38) in an event-triggered control framework, the controller gain K_x can be chosen by using the pole assignment method simply.*

Define $e(k) = [e_y^T(k),\ e_\varpi^T(k)]^T$, with $e_y(k) = y(k) - y(k_i)$, $e_\varpi(k) = \varpi(k) - \varpi(k_i)$, and $\bar{\xi}(k) = [y^T(k),\ \varpi^T(k)]^T$ and updating instants k_i, $i = 0, 1, 2, ...$, are determined by

$$k_{i+1} = \sup\left\{k > k_i \left| \bar{e}^T(k)\bar{\Theta}\bar{e}(k) \le \varepsilon + \varrho\bar{\xi}^T(k)\bar{\Theta}\bar{\xi}(k)\right.\right\}, k \in \mathbb{N}. \tag{12.48}$$

It can be verified that

$$\begin{bmatrix} y(k) \\ \hat{\nu}(k) \end{bmatrix} - \begin{bmatrix} y(k_i) \\ \hat{\nu}(k_i) \end{bmatrix} = L\bar{e}(k),$$

where $L \triangleq \begin{bmatrix} I_q & 0 \\ L_5 & L_4 \end{bmatrix}$,

By applying the event-triggered controller (12.47) to system (12.1), the following closed-loop system can be obtained,

$$\begin{aligned} x(k+1) &= Ax(k) + B(K_c y(k_i) + K_\nu \hat{\nu}(k_i)) + B_d d(k) \\ &= Ax(k) + B(K_c y(k) + K_\nu \hat{\nu}(k)) - BKL\bar{e}(k) + B_d d(k), \\ &= Ax(k) + BK_c y(k) - BKL\bar{e}(k) + BK_\nu H\bar{x}(k) - BK_\nu\epsilon(k) + B_d d(k) \\ &= (A + BK_x)x(k) - BKL\bar{e}(k) - BK_\nu\epsilon(k) + (B_d + BK_d)d(k). \end{aligned} \tag{12.49}$$

Defining $\eta(k) = [\epsilon^T(k),\ x^T(k)]^T$, and with (12.44) we have

$$\eta(k+1) = \tilde{\mathcal{A}}\eta(k) + \tilde{\mathcal{B}}e(k) + \tilde{\mathcal{G}}h_d(k), \tag{12.50}$$

where $h_d(k) = [d^T(k),\ h^T(k)]^T$, and

$$\tilde{\mathcal{A}} = \begin{bmatrix} L_1 & 0 \\ -BK_\nu & A + BK_x \end{bmatrix}, \tilde{\mathcal{B}} = \begin{bmatrix} 0 \\ -BKL \end{bmatrix}, \tilde{\mathcal{G}} = \begin{bmatrix} 0 & \bar{H}E \\ BK_d + B_d & 0 \end{bmatrix}.$$

Here, K_x can be chosen to guarantee that $A + BK_x$ is Schur.

The stability analysis of the closed-loop system (12.50) is given in the following theorem.

Theorem 12.6. *For system (12.50), if there exists a symmetric matrix $\mathcal{P} > 0$ satisfying*

$$\tilde{\mathcal{A}}^T\mathcal{P}\tilde{\mathcal{A}} - \mathcal{P} + \tilde{\mathcal{A}}^T\mathcal{P}\tilde{\mathcal{B}}\tilde{\mathcal{B}}^T\mathcal{P}\tilde{\mathcal{A}} + \tilde{\mathcal{A}}^T\mathcal{P}\tilde{\mathcal{G}}\tilde{\mathcal{G}}^T\mathcal{P}\tilde{\mathcal{A}} + \varrho(\Sigma_{11} + I_{2(n+p)-q}) \le 0, \tag{12.51}$$

where $\Sigma_{11} = \tilde{C}_1^T\Theta\tilde{C}_1$ with $\tilde{C}_1 = \begin{bmatrix} 0 & C \\ -L_4^{-1} & \tilde{L}_{41}H_0 - L_4^{-1}L_5C \end{bmatrix}$, then the states of system (12.50) are globally UUB and the weighting matrix $\bar{\Theta}$ in (12.48) is designed as $\bar{\Theta} = 2I_{n+2p} + \tilde{\mathcal{B}}^T\mathcal{P}\tilde{\mathcal{B}}$.

Proof. The proof is similar to the proof of Theorem 12.1 and the details are omitted here for brevity. □

Remark 12.13. *Considering the similarity of systems (12.49) and (12.18), we can also design the disturbance rejection gain K_d as (12.17).*

It should be noted that, according to the controller (12.47), the system (12.1) is controlled by a constant control input $u(k) \equiv u(k_i)$ for $k \in [k_i, k_{i+1})$, in this situation, due to the unknown external disturbance, the event triggering condition in UIS (12.48) may be frequently satisfied. Thus, for further reducing updating frequency, in what follows, we further propose a predictive event-triggered disturbance rejection control approach by applying the packet-based communication strategy in networked delay compensation control approaches.

In the interval $[k_i, k_{i+1})$, we denote $j \in [0, \Delta_i]$ as the j-th predictive step, where $\Delta_i \geq 1$ denotes the predictive horizon.

Step 1: When $j = 0$, i.e., $k = k_i$, we can directly obtain $u(k_i)$ from (12.47).

Step 2: When $j = 1$, i.e., $k = k_i + 1$, we denote $\tilde{\bar{x}}(k_i) \triangleq [y^T(k_i),\ \hat{\nu}^T(k_i)]^T$, where $y(k_i)$ and $\hat{\nu}(k_i)$ can be obtained from (12.1) and (12.38), respectively. According to (12.3), the one step predictive extended state is generated by

$$\tilde{\bar{x}}(k_i + 1) = \bar{A}\tilde{\bar{x}}(k_i) + \bar{B}u(k_i),$$

and the predictive control input $\tilde{u}(k_i + 1)$ can be designed as $\tilde{u}(k_i + 1) = \mathcal{K}\tilde{\bar{x}}(k_i + 1)$.

Step 3: When $j \geq 2$, i.e., $k \in [k_i + 2, \Delta_i]$, the predictive extended states are determined by

$$\tilde{\bar{x}}(k_i + j) = \bar{A}\tilde{\bar{x}}(k_i + j - 1) + \bar{B}\tilde{u}(k_i + j - 1),$$

and the predictive control input $\tilde{u}(k_i + j)$ can be designed as $\tilde{u}(k_i + j) = \mathcal{K}\tilde{\bar{x}}(k_i + j)$.

For simplicity of analysis, here we denote $\tilde{u}(k_i) \triangleq u(k_i)$, and encapsulate $\tilde{u}(k_i)$, $\tilde{u}(k_i + 1), \cdots, \tilde{u}(k_i + \Delta_i)$ into a packet $U(k_i)$ as

$$U(k_i) = \begin{bmatrix} \tilde{u}(k_i) & \tilde{u}(k_i + 1) & \tilde{u}(k_i + 2) & \cdots & \tilde{u}(k_i + j) & \cdots & \tilde{u}(k_i + \Delta_i) \end{bmatrix}, \quad (12.52)$$

where the predictive horizon Δ_i is determined by

$$\Delta_i = \sup \left\{ j \geq 0 \mid \|\tilde{\bar{x}}(k_i + j) - \tilde{\bar{x}}(k_i)\|^2 \leq \delta_i \right\},$$

and δ_i is a predesigned constant and is used to constrain the size of data packet $U(k_i)$. Moreover, it is easy to obtain that the inequality $\|\sigma(k)\|^2 \leq \delta_i$ holds for $k \in [k_i, k_i + \Delta_i]$, where $\sigma(k) \triangleq \tilde{\bar{x}}(k) - \tilde{\bar{x}}(k_i)$, and it is obvious that $\sigma(k_i) \equiv 0$ for $i \in [1, +\infty)$.

At instants k_i, $i \in [1, +\infty)$, the CIP $U(k_i)$ as shown by (12.52) will be received by the selector. On the one hand, if $k_{i+1} \leq k_i + \Delta_i$, the control input $\tilde{u}(k_i + j)$ in the CIP $U(k_i)$ is selected as the control input of the plant at time instant $k_i + j$, $j \in [1, k_{i+1} - 1]$. On the other hand, if $k_{i+1} > k_i + \Delta_i$, the last predictive control input $\tilde{u}(k_i + \Delta_i)$ in the packet $U(k_i)$ will be used as the control input of the plant for $k \in [k_i + \Delta_i + 1, \; k_{i+1})$. Without loss of generality, we only consider the case that $k_{i+1} > k_i + \Delta_i$, since it includes the case that $k_{i+1} \leq k_i + \Delta_i$.

Under the condition $k_{i+1} > k_i + \Delta_i$, for $k \in [k_i, \; k_i + \Delta_i]$, the closed-loop system is determined as

$$x(k+1) = Ax(k) + B\tilde{u}(k) + B_d d(k)$$

$$= Ax(k) + B\mathcal{K}\bar{x}(k) + B\mathcal{K}(\bar{\tilde{x}}(k) - \bar{\tilde{x}}(k_i)) - B\mathcal{K}\left(\begin{bmatrix} y(k) \\ \hat{\nu}(k) \end{bmatrix} - \bar{\tilde{x}}(k_i) \right)$$

$$- B\mathcal{K}\left(\bar{x}(k) - \begin{bmatrix} y(k) \\ \hat{\nu}(k) \end{bmatrix} \right) + B_d d(k)$$

$$= (A + B\mathcal{K}_x)x(k) + (B_d + B\mathcal{K}_d)d(k)$$

$$+ B\mathcal{K}\sigma(k) - B\mathcal{K}Le(k) - B\mathcal{K}_\nu \epsilon(k). \tag{12.53}$$

Then, similar to (12.50), we have

$$\eta(k+1) = \tilde{A}\eta(k) + \tilde{B}e(k) + \tilde{G}h_d(k) + \tilde{B}^*\sigma(k), \tag{12.54}$$

where $\tilde{B}^* = \begin{bmatrix} 0 \\ B\mathcal{K} \end{bmatrix}$.

For $k \in [k_i + \Delta_i + 1, \; k_{i+1})$, the closed-loop system is

$$x(k+1) = Ax(k) + B\tilde{u}(k_i + \Delta_i) + B_d d(k)$$

$$= (A + B\mathcal{K}_x)x(k) + (B_d + B\mathcal{K}_d)d(k) - B\mathcal{K}Le(k)$$

$$+ B\mathcal{K}\sigma(k_i + \Delta_i) - B\mathcal{K}_\nu \epsilon(k), \tag{12.55}$$

and thus

$$\eta(k+1) = \tilde{A}\eta(k) + \tilde{B}e(k) + \tilde{G}h_d(k) + \tilde{B}^*\sigma(k_i + \Delta_i). \tag{12.56}$$

Above all, the closed-loop system can be given by (12.54) or (12.56) for $k \in \mathbb{I}_{[k_i, \; k_{i+1})}$.

Under the condition $k_{i+1} > k_i + \Delta_i$, the stability analysis of the closed-loop system (12.54) or (12.56) is given in the following theorem and the proof is similar to the proof of Theorem 12.1, which is omitted here.

Theorem 12.7. *For closed-loop system (12.54) or (12.56), if there exists a symmetric matrix $\mathcal{P} > 0$ satisfying*

$$\tilde{A}^T\mathcal{P}\tilde{A} - \mathcal{P} + \tilde{A}^T\mathcal{P}\tilde{B}\tilde{B}^T\mathcal{P}\tilde{A} + \tilde{A}^T\mathcal{P}\tilde{G}^*\tilde{G}^{*T}\mathcal{P}\tilde{A} + \varrho(\Sigma_{11} + I_{2(n+p)-q}) \leq 0, \tag{12.57}$$

where $\tilde{G}^* = \begin{bmatrix} 0 & \bar{H}E & 0 \\ B_d + B\mathcal{K}_d & 0 & B\mathcal{K} \end{bmatrix}$, then the states of system (12.54) or (12.56) are UUB, and the weighting matrix $\bar{\Theta}$ in (12.48) is chosen as $\bar{\Theta} = 2I_{n+2p} + \tilde{B}^T\mathcal{P}\tilde{B}$.

Remark 12.14. *For Theorems 12.6 and 12.7, the conditions (12.51) and (12.57) are different due to $\tilde{G} \neq \tilde{G}^*$, thus the solution $\mathcal{P}$ of (12.51) and (12.57) is also different. Thus, the upper bound of $\|\eta(k)\|$ is different for the proposed two event-triggered control approaches in this section.*

Remark 12.15. *If the predictive horizon is $\Delta_i = 0$, the predictive strategy can be reduced into the previous one, and $\sigma(k) = 0$ holds in $[k_i,\ k_{i+1})$, which leads to $\tilde{\mathcal{G}}^* = \tilde{\mathcal{G}}$, and thus Theorem 12.6 is the special case of Theorem 12.7.*

Remark 12.16. *Motivated by the networked delay compensation control approaches [101], [207], and [179], in this chapter, the predictive event-triggered control approach is proposed. It is worth mentioning that, during the time-interval $(k_i,\ k_i+1)$, instead of the input $u(k_i)$, the predictive control input $\tilde{u}(k_i + j)$ is applied at time step $k_i + j$, $j = 1,\ 2,\cdots,\ k_{i+1} - 1$, which may further reduce the updating times.*

Remark 12.17. *For the proposed predictive event-triggered control approach, by following a similar line of Theorem 12.2, K_d can be designed as (12.17), and thus the bounded disturbance $d(k)$ can be rejected in the output channels at the steady-state.*

12.6   NUMERICAL SIMULATION AND EXPERIMENTS

Consider the following inverted pendulum system [38]:

$$\dot{z}(t) = \begin{bmatrix} 0 & 1 \\ \frac{3(M+m)g}{l(4M+m)} & 0 \end{bmatrix} z(t) + \begin{bmatrix} 0 \\ -\frac{3}{l(4M+m)} \end{bmatrix} u(t), \tag{12.58}$$

where $l = 0.5$ m, $M = 8.0$ kg, $m = 2.0$ kg, and $g = 9.8$ m/s^2. Suppose the sampling period is $T_s = 0.1$ s, then the discretized system of (12.58) can be determined by (12.1) with

$$A = \begin{bmatrix} 1.0877 & 0.1029 \\ 1.7797 & 1.0877 \end{bmatrix}, \ B = \begin{bmatrix} -0.0009 \\ -0.0182 \end{bmatrix}, B_d = \begin{bmatrix} 0.2 \\ 0.8 \end{bmatrix}, \ C = \begin{bmatrix} 1 & 0 \end{bmatrix}.$$

The disturbance in the system (12.1) is assumed to be $d(k) = 10\sin(0.01k\pi)$.

By taking the eigenvalues of matrices $A+BK_x$ and $\bar{A}-L\bar{C}$ as 0.25, 0.26, and 0.25, 0.26, 0.27, 0.28, respectively, and by applying the well-known pole assignment technique, the gain matrices K_x and L can be calculated as $K_x = \begin{bmatrix} 408.0064 & 71.5977 \end{bmatrix}$ and $L = \begin{bmatrix} 3.1154 & 15.2364 & 6.1066 & 10.6095 \end{bmatrix}^T$, respectively. According to the Theorem 12.2, it is obvious that rank condition (12.16) is satisfied and it follows from (12.17) that the disturbance compensation gain K_d is determined as $K_d = 152.7990$. The weighting-matrix Θ is derived as

$$\Theta = \begin{bmatrix} 28.7036 & 5.037 & 0 & 10.7495 \\ 5.037 & 0.8839 & 0 & 1.8863 \\ 0 & 0 & 0 & 0 \\ 10.7495 & 1.8863 & 0 & 4.0257 \end{bmatrix}.$$

Furthermore, it can be verified that the condition (12.13) is feasible.

For given parameters $\varepsilon = 0.01$ and $\varrho = 0.001$ in (12.6), simulation results are displayed in Figure 12.1.

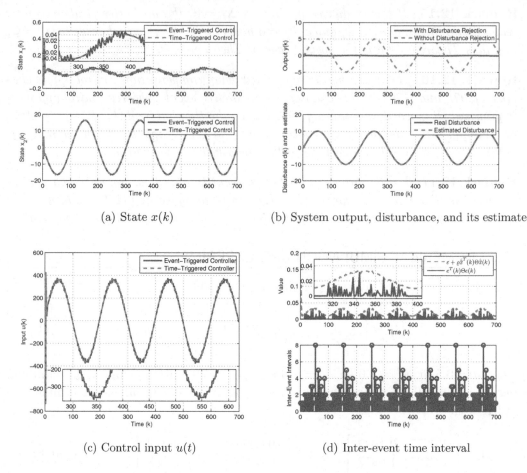

(a) State $x(k)$

(b) System output, disturbance, and its estimate

(c) Control input $u(t)$

(d) Inter-event time interval

Figure 12.1 Comparative simulation results of event-triggered disturbance rejection control approach (full-order observer case)

The curves of state $x(k)$ and control input $u(t)$ of the system with event-triggered controller (12.5) and conventional time-triggered controller

$$u(k) = K\hat{\bar{x}}(k) = K_x\hat{x}(k) + K_d\hat{d}(k) \tag{12.59}$$

are plotted in Figure 12.1(a) and Figure 12.1(c), respectively. It is clear that with the presented event-triggered controller (12.5), the states are UUB. Moreover, the performance of the proposed event-triggered control system is close to that of the conventional discrete-time control system [with the time-triggered controller (12.59)].

To show the active disturbance rejection, the curves of output $y(k)$ of the system with event-triggered disturbance rejection controller (12.5) and

$$u(k) = K_x\hat{x}(k_i), k \in [k_i, k_{i+1}) \tag{12.60}$$

are plotted in Figure 12.1(b). It is clear that, at the steady-state, the disturbance is actively rejected in the output channel. In addition, the disturbance and its estimate are also shown in Figure 12.1(b).

Moreover, the inter-event intervals are plotted in Figure 12.1(d). For the system with the time-triggered controller (12.59), 700 updating instants of the controller are needed. However, there are only 412 controller updating instants for the proposed event-triggered control strategy. It is easy to conclude that the controller updating frequency is remarkably reduced by applying the proposed event-triggered control strategy, and thus the burden of the processor is lessened sharply.

Next, we further examine the effectiveness of the reduced-order ESO-based event-triggered control approach, and assume that $B_d = 3.5B$. By using the pole assignment technique, and selecting the eigenvalues of matrix $\bar{A}_{22} + L_r \bar{A}_{12}$ as 0.25, 0.39, 0.45, we can compute the observer gain as $L_r = \begin{bmatrix} -14.1669 & 132.1517 & 172.3089 \end{bmatrix}^T$. Then, we choose the transform matrix

$$T = \begin{bmatrix} 1 & 0 & 0 & 0 \\ -14.1669 & 1 & 0 & 0 \\ 132.1517 & 0 & 1 & 0 \\ 172.3089 & 0 & 0 & 1 \end{bmatrix},$$

and select the eigenvalues of matrix $(\tilde{A}_{11} + (\tilde{A}_{12} + \tilde{B}_1 K_{rd})T_3 T_1^{-1}) + \tilde{B}_1 K_r$ as 0.32 and 0.33 to obtain the controller gain $K_r = \begin{bmatrix} 1899.4 & 66.6 \end{bmatrix}$. The weighting-matrix Θ_r is derived as

$$\Theta_r = 10^5 \times \begin{bmatrix} 1.5241 & 0.0535 & 0 & -0.0028 \\ 0.0535 & 0.0019 & 0 & -0.0001 \\ 0 & 0 & 0 & 0 \\ -0.0028 & -0.0001 & 0 & 0 \end{bmatrix}.$$

By resorting to Matlab Control Toolbox, we can verify that condition (12.29) is feasible.

Taking $\bar{\varepsilon} = 0.01$ and $\bar{\varrho} = 0.01$ in the UIS (12.26), we obtain the simulation results that are plotted in Figure 12.2. By applying the proposed event-triggered controller and traditional time-triggered controller with disturbance rejection, the states and control inputs are shown in Figure 12.2(a) and Figure 12.2(c), respectively. It is clear from Figure 12.2(a) that the states of the event-triggered control system are UUB. With the disturbance estimate, the disturbance is rejected in the output channel at the steady-state as displayed in Figure 12.2(b). Moreover, it should be pointed out that the traditional discrete-time system requires 700 updating instants, whereas our proposed event-triggered control scheme needs only 319 updating instants to achieve almost the same performance. The inter-event intervals are shown in Figure 12.2(d). Thus, it can be concluded that with the proposed event-triggered controller, not only the updating frequency of the controller can be considerably reduced, but also the disturbance can be rejected efficiently.

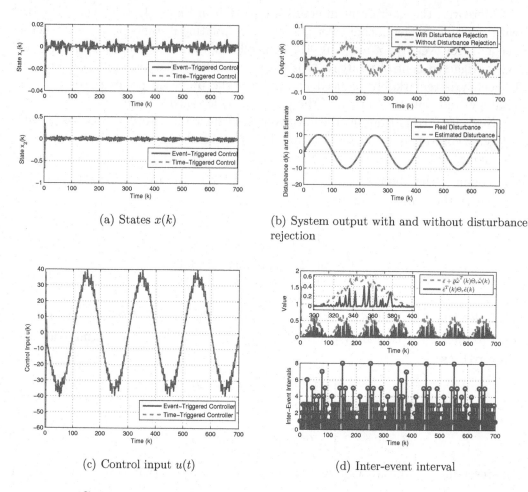

(a) States $x(k)$

(b) System output with and without disturbance rejection

(c) Control input $u(t)$

(d) Inter-event interval

Figure 12.2 Comparative simulation results of event-triggered disturbance rejection control approach (reduced-order ESO case)

In what follows, we will examine the validity the proposed reduced-order EFSO-based event-triggered control approach. Firstly, we assume the disturbance $d(k)$ is matched and $B_d = 50B$. By placing the poles of matrix $\bar{A}_{22} - L_5\bar{A}_{12}$ as 0.25, 0.26, and 0.27, we obtain that $L_5 = \begin{bmatrix} 15.4083 & -11.5314 & -16.0485 \end{bmatrix}^T$ and by taking $L_4 = 40I_3$, the other parameters of RESFO can be calculated according to (12.39)–(12.41),

$$L_1 = \begin{bmatrix} -0.4978 & 0 & -0.2166 \\ 1.1866 & 0 & 0.4811 \\ 1.6514 & -1 & 1.2778 \end{bmatrix}, L_2 = \begin{bmatrix} -0.4793 & 0.5776 & 0.8481 \end{bmatrix}^T,$$

$$L_3 = \begin{bmatrix} -0.1083 & -0.2595 & -0.3611 \end{bmatrix}^T \times 10^{-3}.$$

By taking the poles of matrix $A + B\mathcal{K}_x$ as 0.25 and 0.26, the controller gain is calculated as $\mathcal{K}_x = [445.4240, 74.4241]$, and the disturbance rejection gain can be determined as $K_d = -50$. For given parameters $\varepsilon = 0.01$ and $\varrho = 0.001$, it can be verified that both the conditions (12.51) and (12.57) are feasible, and the weighting matrix $\bar{\Theta}$ in (12.48) is

$$\bar{\Theta} = \begin{bmatrix} 16.2936 & 17.7698 & 0 & -11.9382 \\ 17.7698 & 24.0914 & 0 & -14.8416 \\ 0 & 0 & 2 & 0 \\ -11.9382 & -14.8416 & 0 & 11.9710 \end{bmatrix}.$$

Then, it can be found in Figure 12.3(a) that, compared with [139], the unknown disturbance $d(k)$ can be estimated with the proposed reduced-order EFSO more accurately. From Figure 12.3(b), we can find that, by using the proposed two methods in Section 12.5, the external disturbance can be rejected in the output channel at the steady-state. Moreover, the control input signals are plotted in Figure 12.3(c) and the inter-event intervals of the system with event-triggered control and predictive event-triggered control approaches proposed in Section 12.5 is given in Figure 12.3(d), and it is clear that, by applying the predictive event-triggered control approach, not only the updating times are reduced, but also the control performance is improved.

Table 12.1 Updating times of controller for different Δ_i (matched disturbance case)

	$\Delta_i = 0$ (Controller (12.47))	$\Delta_i = 1$	$\Delta_i = 2$	$\Delta_i = 3$
Updating times	362	220	185	154

As shown in Table 12.1, there are 362 updating times for event-triggered controller (12.47), and by applying the proposed predictive event-triggered control scheme, only 220, 185, and 154 event triggering times are needed for different predictive horizons $\Delta_i = 1$, $\Delta_i = 2$, and $\Delta_i = 3$, $i \in [1, +\infty)$, respectively.

Now, we consider that the disturbance $d(k)$ is mismatched, and $B_d = [0.2, 0.8]^T$. By placing the poles of matrix $\bar{A}_{22} - L_5 \bar{A}_{12}$ as 0.25, 0.26, and 0.27, L_5 can be determined as $L_5 = \begin{bmatrix} 10.6697 & -0.2053 & -6.0489 \end{bmatrix}^T$ and taking $L_4 = 40I_3$, the other parameters of reduced-order EFSO are obtained as

$$L_1 = \begin{bmatrix} -0.0102 & 0 & -1.3339 \\ 0.0211 & 0 & 1.0411 \\ -0.6224 & -1 & 0.7902 \end{bmatrix}, L_2 = \begin{bmatrix} -0.4501 & 0.1687 & -0.2059 \end{bmatrix}^T,$$

$$L_3 = \begin{bmatrix} -0.2149 & -0.0046 & -0.1361 \end{bmatrix}^T \times 10^{-3}.$$

Taking the poles of matrix $A + BK_x$ as 0.25, 0.2, the controller gain is computed as $K = [406.9687, 71.3807, 0, -152.3046]$. Given $\varepsilon = 0.01$ and $\varrho = 0.001$, both (12.51) and (12.57) can be verified to be feasible, and the weighting matrix $\bar{\Theta}$ can be determined as

$$\bar{\Theta} = \begin{bmatrix} 11.2394 & 12.6231 & 0 & 26.9339 \\ 12.6231 & 19.2461 & 0 & 36.7979 \\ 0 & 0 & 2 & 0 \\ 26.9339 & 36.7979 & 0 & 80.5155 \end{bmatrix}.$$

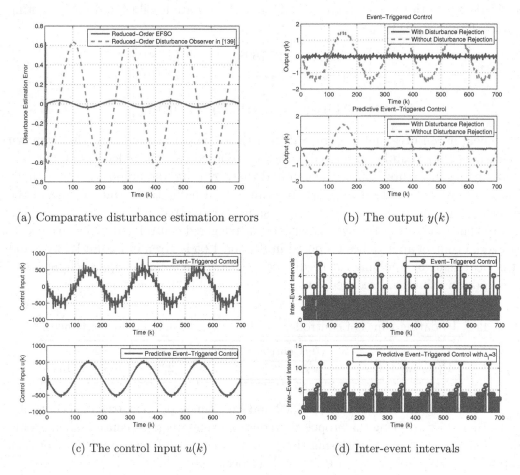

(a) Comparative disturbance estimation errors

(b) The output $y(k)$

(c) The control input $u(k)$

(d) Inter-event intervals

Figure 12.3 Simulation results of the reduced-order EFSO-based event-triggered control approach (matched disturbance case)

As shown by Figure 12.4(a), the unknown disturbance $d(k)$ can be estimated more accurately by using the proposed reduced-order EFSO, and Figure 12.4(b) shows that the external disturbance can be always rejected in the output channels at the steady-state. In Figure 12.4(c), the control input signals of system are also presented. In Figure 12.4(d), the inter-event interval for both the event-triggered control and predictive event-triggered control approaches proposed in Section 12.5 are presented. As given in Table 12.2, there are 390 updating instants that are required for system with the event-triggered controller (12.47), and with the proposed predictive event-triggered control approach, only 203, 171, and 149 updating instants are needed for the predictive horizons are $\Delta_i = 1$, $\Delta_i = 2$, and $\Delta_i = 3$, $i \in [1, +\infty)$, respectively.

Finally, we also apply the proposed reduced-order EFSO-based event-triggered control approach to the rotary inverted pendulum system, and the experimental platform of rotary inverted pendulum is shown in Figure 12.5(a).

The schematic diagram of rotary inverted pendulum is given in Figure 12.5(b), where L_r and L_p denote the length of arm and pendulum, respectively. J_r and J_p

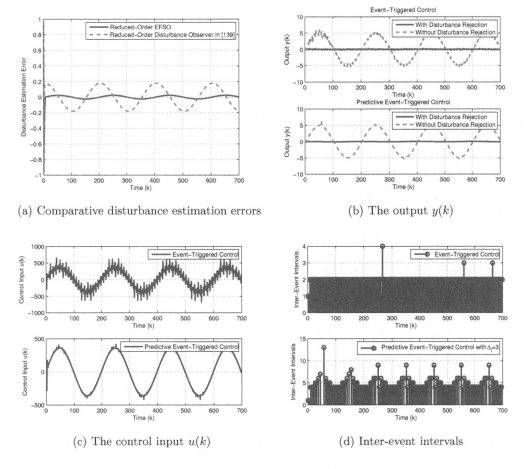

(a) Comparative disturbance estimation errors

(b) The output $y(k)$

(c) The control input $u(k)$

(d) Inter-event intervals

Figure 12.4 Simulation results of the system with mismatched disturbance $f(k)$

denote the inertia moment of arm and pendulum, respectively. θ and α denote the angle of arm and pendulum, respectively. m_p and g are the mass of pendulum and gravitational acceleration, respectively. Define the state $x(t) = [\theta(t),\ \alpha(t),\ \dot{\theta}(t),\ \dot{\alpha}(t)]^T$, we can obtain the linearized system model as

$$\dot{x}(t) = \begin{bmatrix} 0_2 & I_2 \\ R_1 & R_2 \end{bmatrix} x(t) + \begin{bmatrix} 0_{2\times 1} \\ R_3 \end{bmatrix} u(t),$$

$$y(t) = \begin{bmatrix} I_2 & 0_2 \end{bmatrix} x(t),$$

Table 12.2 Updating times of controller for different Δ_i (mismatched disturbance case)

	$\Delta_i = 0$ (Controller (12.47))	$\Delta_i = 1$	$\Delta_i = 2$	$\Delta_i = 3$
Updating times	390	203	171	149

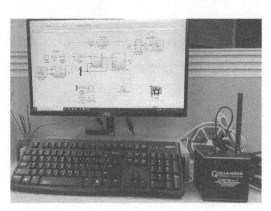

(a) The rotary inverted pendulum

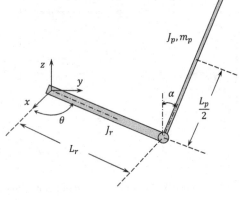

(b) Schematic diagram

Figure 12.5 The rotary inverted pendulum and its schematic diagram

where

$$R_1 = \begin{bmatrix} 0 & \frac{m_p^2 L_p^2 L_r g}{4J_T} \\ 0 & \frac{m_p L_p g(J_r + m_p L_r^2)}{2J_T} \end{bmatrix}, R_2 = \begin{bmatrix} -\frac{(4J_p + m_p L_p^2)D_r}{4J_T} & -\frac{m_p L_p L_r D_p}{2J_T} \\ -\frac{m_p L_p L_r D_r}{2J_T} & -\frac{(J_r + m_p L_r^2)D_p}{J_T} \end{bmatrix},$$

$$R_3 = \begin{bmatrix} \frac{4J_p + m_p L_p^2}{4J_T} \\ -\frac{m_p L_p L_r}{2J_T} \end{bmatrix}.$$

Taking the sampling period as 0.002s, the discretized system matrices are given by

$$A = \begin{bmatrix} 1 & 0.0003 & 0.002 & 0 \\ 0 & 1.005 & 0 & 0.002 \\ 0 & 0.2986 & 1 & 0.0003 \\ 0 & 0.5233 & 0 & 1.0005 \end{bmatrix}, B = \begin{bmatrix} 0.0001 \\ 0.0001 \\ 0.0995 \\ 0.0983 \end{bmatrix}.$$

Assume $B_d = 5B$, and there exists a matched disturbance in the system, $d(k) = 0.2\sin\left(\frac{k\pi}{1500}\right)$. By placing the poles of $\bar{A}_{22} - L_5\bar{A}_{12}$ as 0.52, 0.61, 0.9, and 0.91,

L_5 can be designed as $L_5 = \begin{bmatrix} 124.7104 & 74.8341 & -23.78 & -29.9574 \\ 294.6424 & 340.865 & 269.4026 & 293.214 \end{bmatrix}^T$, and by

taking $L_4 = I_4$, the other parameters of reduced-order EFSO are calculated as

$$L_1 = \begin{bmatrix} 0.7506 & -0.5891 & 0 & 0.2905 \\ -0.1497 & 0.3187 & 0 & 0.2868 \\ 0.0476 & -0.5389 & 0 & 0.8794 \\ 0.0599 & -0.5865 & -1 & 1.8708 \end{bmatrix},$$

$$L_2 = -\begin{bmatrix} 83.8946 & 78.2464 & 36.9605 & 38.7255 \\ 189.0234 & 191.9272 & 181.358 & 196.4947 \end{bmatrix}^T,$$

$$L_3 = \begin{bmatrix} 0.0581 & 0.0574 & -0.0241 & -0.0258 \end{bmatrix}^T.$$

By choosing 0.57, 0.9915, 0.9916, and 0.9917 as the poles of matrix $A + B\mathcal{K}_x$, the controller gain is obtained as $\mathcal{K} = [2.8075, -63.8887, 2.0104, -6.6133, 0, -5]$. Given $\varepsilon = 0.01$ and $\varrho = 3.6 \times 10^{-9}$, the matrix inequality (12.51) can be verified to be feasible, and the weighting matrix $\bar{\Theta}$ can be determined as

$$
\bar{\Theta} = \begin{bmatrix}
3.366 & 0.907 & -0.007 & 0.023 & 0 & -0.018 \\
0.907 & 2.6 & -0.005 & 0.015 & 0 & -0.012 \\
-0.007 & -0.005 & 2 & -0.0001 & 0 & 0.0001 \\
0.023 & 0.015 & -0.0001 & 2.0004 & 0 & -0.0003 \\
0 & 0 & 0 & 0 & 2 & 0 \\
-0.018 & -0.012 & 0.0001 & -0.0003 & 0 & 2.0002
\end{bmatrix}.
$$

By applying the reduced-order EFSO-based event-triggered control approach to the rotary inverted pendulum system, the experiment results are shown in Figure 12.6.

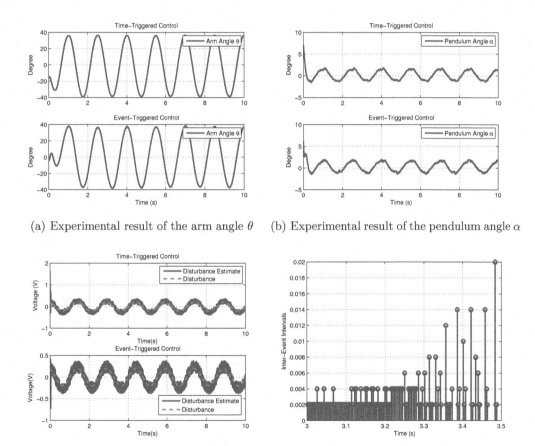

(a) Experimental result of the arm angle θ (b) Experimental result of the pendulum angle α

(c) Experimental result of disturbance estimation   (d) The inter-event intervals in $(3, 3.5)$s

Figure 12.6 Experiment results for the reduced-order EFSO-based event-triggered control approach

It can be seen from Figure 12.6(c) that the disturbance $d(k)$ is well estimated by using the proposed reduced-order EFSO. Noting that there exist 10000 sampling steps in time interval $[0, 20]$s, only 6934 updating times are needed. Figure 12.6(d) presents the inter-event interval in $(3, 3.5)$s. Then, by comparing the outputs of rotary inverted pendulum system in Figure 12.6(a) and Figure 12.6(b), it can be seen that the proposed event-triggered control strategy is effective.

12.7 SUMMARY

This chapter has investigated the event-triggered disturbance rejection control problem for discrete-time systems. The full-order ESO, reduced-order ESO, and EFSO have been designed to determine the estimates of the disturbance and system states. With the estimated disturbance and states, the UIS have been designed, and the event-triggered disturbance rejection controllers also have been designed. Based on the analysis on stability and disturbance rejection performance, it has been shown that the states are UUB and the disturbance is also rejected in the output channel at the steady-state. Finally, numerical simulations and experimental results have been presented to demonstrate the applicability of the developed approaches.

Bibliography

[1] K. Abidi, J.-X. Xu, and X. Yu. On the discrete-time integral sliding-mode control. *IEEE Trans. Autom. Control*, 52(4):709–715, 2007.

[2] J. Almeida, C. Silvestre, and A. M. Pascoal. Self-triggered output feedback control of linear plants in the presence of unknown disturbances. *IEEE Trans. Autom. Control*, 59(11):3040–3045, 2014.

[3] J. Almeida, C. Silvestre, and A. M. Pascoal. Self-triggered state-feedback control of linear plants under bounded disturbances. *Int. J. Robust and Nonlinear Control*, 25(8):1230–1246, 2014.

[4] A. Anta and P. Tabuada. To sample or not to sample: Self-triggered control for nonlinear systems. *IEEE Trans. Autom. Control*, 55(9):2030–2042, 2010.

[5] P. Antsaklis and J. Baillieul. Guest editorial. special issue on networked control systems. *IEEE Trans. Autom. Control*, 49(9):1421–1423, 2004.

[6] K.-E. Arzen. A simple event-based PID controller. In *Proc. 14th IFAC World Congress*, pages 423–428, Beijing, 1999.

[7] D. P. Borgers, V. S. Dolk, and W. P. M. H. Heemels. Distributed dynamic event-triggered control for multi-agent systems. In *Proc. of IEEE 55th Annual Conference on Decision and Control*, pages 1352–1357, Las Vegas, USA, 2016.

[8] D. P. Borgers, V. S. Dolk, and W. P. M. H. Heemels. Riccati-based design of event-triggered controllers for linear systems with delays. *IEEE Trans. Autom. Control*, 63(1):174–188, 2018.

[9] R. W. Brockett and D. Liberzon. Quantised feedback stabilization of linear systems. *IEEE Trans. Autom. Control*, 45(7):1279–1289, 2000.

[10] X.-H. Chang and G.-H. Yang. Delay-dependent stabilization conditions of controlled positive T-S fuzzy systems with time varying delay. *Int. J. of Innovative Computing, Information and Control*, 7(4):1533–1548, 2011.

[11] X.-H. Chang and G.-H. Yang. Relaxed results on stabilization and state feedback H_∞ control conditions for T-S fuzzy systems. *Int. J. of Innovative Computing, Information and Control*, 7(4):1753–1764, 2011.

[12] W.-W. Che and G.-H. Yang. Quantised H_∞ filter design for discrete-time systems. *Int. J. Control*, 82(2):195–206, 2009.

[13] B. Chen, Y. Niu, and Y. Zou. Sliding mode control for stochastic markovian jumping systems with incomplete transition rate. *IET Control Theory Appl.*, 7(10):1330–1338, 2013.

[14] B. Chen, Y. Niu, and Y. Zou. Sliding mode control for stochastic markovian jumping systems subject to successive packet losses. *J. Frank. Inst.-Eng. Appl. Math.*, 351:2169–2184, 2014.

[15] T. Chen and B. A. Francis. *Optimal Sampled-Data Control Systems.* Springer, London, U.K., 1995.

[16] W.-H. Chen and W. X. Zheng. An improved stabilization method for sampled-data control systems with control packet loss. *IEEE Trans. Autom. Control*, 59(9):2378–2384, 2012.

[17] O. L.V. Costa, M. D. Fragoso, and R. P. Marques. *Discrete-time Markov jump linear systems.* Springer-Verlag, New York, 2005.

[18] D. B. Dacic and D. Nesic. Quadratic stabilization of linear networked control systems via simultaneous protocol and controller design. *Automatica*, 43:1145–1155, 2007.

[19] D. F. Delchamps. Stabilizing a linear system with quantized state feedback. *IEEE Trans. Autom. Control*, 35(8):916–924, 1990.

[20] V. S. Dolk, D. P. Borgers, and W. P. M. H. Heemels. Output-based and decentralized dynamic event-triggered control with guaranteed l_p-gain performance and zeno-freeness. *IEEE Trans. Autom. Control*, 62(1):34–49, 2017.

[21] H. Dong, Z. Wang, D. W. C. Ho, and H. Gao. Robust H_∞ fuzzy output-feedback control with multiple probabilistic delays and multiple missing measurements. *IEEE Trans. Fuzzy Syst.*, 18(4):712–725, 2010.

[22] H. Dong, Z. Wang, J. Lam, and H. Gao. Fuzzy-model-based robust fault detection with stochastic mixed time delays and successive packet dropouts. *IEEE Trans. Syst., Man Cybern. B, Cybern.*, 42(2):365–376, 2012.

[23] M. C. F. Donkers and W. P. M. H. Heemels. Output-based event-triggered control with guaranteed $\mathcal{L}_\infty$-gain and improved and decentralised event-triggering. *IEEE Trans. Autom. Control*, 57(6):1362–1376, 2012.

[24] M. C. F. Donkers, W. P. M. H. Heemels, Nathan van de Wouw, and Laurentiu Hetel. Stability analysis of networked control systems using a switched linear systems approach. *IEEE Trans. Autom. Control*, 56(9):2101–2115, 2011.

[25] N. Elia and S. K. Mitter. Stabilization of linear systems with limited information. *IEEE Trans. Autom. Control*, 46(9):1384–1400, 2001.

[26] A. Eqtami, D. V. Dimarogonas, and K. J. Kyriakopoulos. Event-triggered control for discrete-time systems. In *Proc. American Control Conference*, pages 4719–4724, Marriott Waterfront, Baltimore, MD, USA, 2010.

[27] F. Fagnani and S. Zampieri. Stability analysis and synthesis for scalar linear systems with a quantized feedback. *IEEE Trans. Autom. Control*, 48(9):1569–1584, 2003.

[28] Y. Fan, L. Liu, G.ang Feng, and Y. Wang. Self-triggered consensus for multi-agent systems with zeno-free triggers. *IEEE Trans. Autom. Control*, 60(10):2779–2784, 2015.

[29] Y. Fang and K. A. Loparo. Stochastic stability of jump linear systems. *IEEE Trans. Autom. Control*, 47(7):1204–1208, 2002.

[30] G. Feng. A survey on analysis and design of model-based fuzzy control systems. *IEEE Trans. Fuzzy Syst.*, 14(5):676–697, 2006.

[31] T. Fernando and H. Trinh. Design of reduced-order state/unknown input observers based on a descriptor system approach. *Asian J. Control*, 9(4):458–465, 2007.

[32] G. Franklin, J. Powell, and M. Workman. *Digital Control of Dynamic Systems*. NJ:Prentice-Hall, Englewood Cliffs, 1997.

[33] E. Fridman. A refined input delay approach to sampled-data control. *Automatica*, 46(2):421–427, 2010.

[34] E. Fridman and U. Shaked. An improved stabilization method for linear time-delay systems. *IEEE Trans. Automat. Control*, 47:1931–1937, 2002.

[35] M. Fu and L. Xie. The sector bound approach to quantized feedback control. *IEEE Trans. Autom. Control*, 50(11):1698–1711, 2005.

[36] K. Fukuda, K. Fujita, and T. Ushio. Dynamic event-triggered minimal-order observer for linear systems. In *Proc. of the Second International Conference on Event-based Control, Communication, and Signal Processing*, pages 1–8, Krakow, Poland, 2016.

[37] P. Gahinet and P. Apkarian. A linear matrix inequality approach to H_∞ control. *Int. J. Robust Nonlinear Control*, 4(4):421–448, 1994.

[38] H. Gao and T. Chen. New results on stability of discrete-time systems with time-varying state delay. *IEEE Trans. Autom. Control*, 52(1):328–334, 2007.

[39] H. Gao and T. Chen. Networked-based H_∞ output tacking control. *IEEE Trans. Autom. Control*, 53(9):2142–2148, 2008.

[40] H. Gao and T. Chen. A new approach to quantized feedback control systems. *Automatica*, 44:534–542, 2008.

[41] H. Gao, T. Chen, and T. Chai. Passivity and passification for networked control systems. *SIAM J. Control and Optimization*, 46(4):1299–1322, 2007.

[42] H. Gao, T. Chen, and J. Lam. A new delay system approach to network-based control. *Automatica*, 44:39–52, 2008.

[43] H. Gao, J. Lam, C. Wang, and Y. Wang. Delay-dependent output feedback stabilization of discrete-time systems with time-varying state delay. *IEE Proc.-Control Theory Appl.*, 151(6):691–698, 2004.

[44] H. Gao, X. Meng, and T. Chen. Stabilization of networked control systems with a new delay characterization. *IEEE Trans. Automat. Control*, 53(9):2142–2148, 2008.

[45] H. Gao, X. Meng, T. Chen, and J. Lam. Stabilization of networked control systems via dynamic output-feedback controllers. *SIAM J. Control Optim.*, 48(5):3643–3658, 2010.

[46] H. Gao, J. Wu, and P. Shi. Robust sampled-data H_∞ control with stochastic sampling. *Automatica*, 45(7):1729–1736, 2009.

[47] E. Garcia and P. J. Antsaklis. Model-based event-triggered control for systems with quantization and time-varying network delays. *IEEE Trans. Autom. Control*, 58(2):422–434, 2013.

[48] X. Ge, Q.-L. Han, and Z. Wang. A dynamic event-triggered transmission scheme for distributed set-membership estimation over wireless sensor networks. *IEEE Trans. Cybern.*, 49(1):171–183, 2019.

[49] J. C. Geromel, C. E. de Souza, and R. E. Skelton. Static output feedback controllers: stability and convexity. *IEEE Trans. Autom. Control*, 43(1):120–125, 1998.

[50] A. Gersho and R. M. Gray. *Vector Quantization and Signal Compression*. Kluwer Academic Publishers, New York, 1992.

[51] A. Girard. Dynamic triggering mechanisms of event-triggered control. *IEEE Trans. Autom. Control*, 60(7):1992–1997, 2015.

[52] L. Grüne, F. Müller, S. Jerg, O. Junge, M. Post, D. Lehmann, and J. Lunze. Two complementary approaches to event-based control. *Automatisierungstechnik*, 58:173–182, 2010.

[53] Z. Gu, Z. Huan, D. Yue, and F. Yang. Event-triggered dynamic output feedback control for networked control systems with probabilistic nonlinearities. *Information Sciences*, 457–458:99–112, 2018.

[54] B.-Z. Guo and Z.-L. Zhao. *Active disturbance rejection control for nonlinear systems : an introduction*. John Wiley & Sons, Singapore, 2016.

[55] Y. Guo and S. Li. A new networked predictive control approach for systems with random network delay in the forward channel. *Int. J. Syst. Sci.*, 41(5):511–520, 2010.

[56] R. Gupta and M. Chow. Networked control system: overview and research trends. *IEEE Trans. Ind. Electron.*, 57(7):2527–2535, 2010.

[57] Q. P. Ha, N. D. That, P. T. Nam, and H. Trinh. Partial state estimation for linear systems with output and input time delays. *ISA Trans.*, 53(2):327–334, 2014.

[58] Q. P. Ha and H. Trinh. State and input simultaneous estimation for a class of nonlinear systems. *Automatica*, 40(10):1779–1785, 2004.

[59] J. Han. From PID to active disturbance rejection control. *IEEE Trans. Ind. Electron.*, 56(3):900–906,, 2009.

[60] J. Hauser, S. Sastry, and P. Kokotović. Nonlinear control via approximate input-output linearization: the ball and beam example. *IEEE Trans. Automat. Control*, 37(3):392–398, 1992.

[61] T. Hayakawa, H. Ishii, and K. Tsumura. Adaptive quantized control for linear uncertain discrete-time systems. *Automatica*, 45(3):692–700, 2009.

[62] T. Hayakawa, H. Ishii, and K. Tsumura. Adaptive quantized control for nonlinear uncertain discrete-time systems. *Systems & Control Letters*, 58(9):625–632, 2009.

[63] X. He, Z. Wang, Y. D. Ji, and D. H. Zhou. Network-based fault detection for discrete-time state-delay systems: A new measurement model. *International Journal of Adaptive Control and Signal Processing*, 22(5):510–528, 2007.

[64] X. He, Z. Wang, Y. Liu, and D. Zhou. Least-squares fault detection and diagnosis for networked sensing systems using direct state estimation approach. *IEEE Trans. Ind. Informatics*, 9(3):1670–1679, 2013.

[65] X. He, Z. Wang, X. Wang, and D. Zhou. Networked strong tracking filtering with multiple packet dropouts: algorithms and applications. *IEEE Trans. Ind. Electron.*, 61(3):1454–1463, 2014.

[66] Y. He, M. Wu, G.-P. Liu, and J.-H. She. Output feedback stabilization for a discrete-time system with a time-varying delay. *IEEE Trans. Automat. Control*, 53(10):2372–2377, 2008.

[67] W. P. M. H. Heemels and M. C. F. Donkers. Model-based periodic event-triggered control for linear systems. *Automatica*, 49(3):698–711, 2013.

[68] W. P. M. H. Heemels, M. C. F. Donkers, and A. R. Teel. Periodic event-triggered control based on state feedback. In *Proc. 50th IEEE Conference on Decision and Control/European Control Conference*, pages 2571–2576, 2011.

[69] W. P. M. H. Heemels, M. C. F. Donkers, and A. R. Teel. Periodic event-triggered control for linear systems. *IEEE Trans. Automat. Control*, 58(4):847–861, 2013.

[70] W. P. M. H. Heemels, R. Gorter, A. van Zijl, P. van den Bosch, S. Weiland, W. Hendrix, and M. Vonder. Asynchronous measurement and control: a case study on motor synchronization. *Contr. Eng. Pract.*, 7(12):1467–1482, 1999.

[71] J. P. Hespanha, P. Naghshtabrizi, and Y. Xu. A survey of recent results in networked control systems. *Proc. of IEEE*, 95(1):138–162, 2007.

[72] L. Hetel, J. Daafouz, and C. Lung. Analysis and control of lti and switched systems in digital loops via an event-based modelling. *Int. J. Control*, 81(7):1125–1138, 2008.

[73] L.-S. Hu, T. Bai, P. Shi, and Z. Wu. Sampled-data control of networked linear control systems. *Automatica*, 43:903–911, 2007.

[74] S. Hu and D. Yue. L_2-gain analysis of event-triggered networked control systems: a discontinuous lyapunov functional approach. *Int. J. Robust and Nonlinear Control*, 23(11):1277–1300, 2013.

[75] S. Hu, D. Yue, and J. Liu. H_∞ filtering for networked systems with partly known distribution transmission delays. *Information Sciences*, 194:270–282, 2012.

[76] S. Hu, D. Yue, X. Yin, X. Xie, and Y. Ma. Adaptive event-triggered control for nonlinear discrete-time systems. *Int. J. Robust Nonlinear Control*, 26(8):4104–4125, 2016.

[77] W. Hu, G.-P. Liu, and D. Rees. Event-driven networked predictive control. *IEEE Trans. Ind. Electron.*, 54(3):1603–1613, 2007.

[78] D. Huang and S. K. Nguang. *Robust Control for Uncertain Networked Control Systems with Random Delays*. Springer-Verlag, Berlin Heidelberg, 2009.

[79] D. Huang and S. K. Nguang. Robust fault estimator design for uncertain networked control systems with random time delays: An ILMI approach. *Information Sciences*, 180(3):465–480, 2010.

[80] Y. Huang, J. Wang, D. Shi, and L. Shi. Toward event-triggered extended state observer. *IEEE Trans. Autom. Control*, 63(6):1842–1849, 2018.

[81] O. C. Imer, S. Yuksel, and T. Basar. Optimal control of LTI systems over unreliable communication links. *Automatica*, 42(9):1429–1439, 2006.

[82] T. Iwasaki and R.E. Skelton. All controllers for the general H_∞ control problems: LMI existence conditions and state space formulas. *Automatica*, 30(8):1307–1317, 1994.

[83] L. Jetto and V. Orsini. A new event-driven output-based discrete-time control for the sporadic MIMO tracking problem. *Int. J. Robust and Nonlinear Control*, 24(5):859–875, 2014.

[84] Y. Ji and H. J. Chizeck. Jump linear quadratic Gaussian control: steady-state solution and testable conditions. *Control Theory Adv. Tech.*, 6:289–319, 1990.

[85] P. Jia, F. Hao, and H. Yu. Function observer based event-triggered control for linear systems with guaranteed L_∞-gain. *IEEE/CAA J. Automatica Sinica*, 2(4):394–402, 2015.

[86] H. Khalil. *Nonlinear Syst., 3rd ed.* Prentice Hall, Upper Saddle River, NJ, 2002.

[87] A. Kruszewski, W. J. Jiang, E. Fridman, J. P. Richard, and A. Toguyeni. A switched system approach to exponential stabilization through communication network. *IEEE Trans. Control Syst. Technol.*, 20(4):887–900, 2012.

[88] D. Lehmann. *Event-Based State-Feedback Control.* Logos Verlag, 2011.

[89] D. Lehmann and J. Lunze. Event-based control using quantized state information. In *Proc. 2th IFAC Workshop on Disributed Estimation and Control in Networked Systems*, Annecy, France, 2010.

[90] D. Lehmann and J. Lunze. Event-based output-feedback control. In *Proc. 19th Mediterranean Conference on Control and Automation*, pages 982–987, Greece, 2011.

[91] D. Lehmann and J. Lunze. Event-based control with communication delays and packet losses. *Int. J. Control*, 85(5):563–577, 2012.

[92] L. Li and M. Lemmon. Event-triggered output feedback control of finite horizon discrete-time multi-dimensional linear processes. In *49th IEEE Conference on Decision and Control*, pages 3221–3226, Atlanta, 2010.

[93] P. Li, Y. Kang, Y.-B. Zhao, and J. Zhou. Dynamic event-triggered control for networked switched linear systems. In *Proc. of the 36th Chinese Control Conference*, pages 7984–7989, Dalian, China, 2017.

[94] Q. Li, B. Shen, Z. Wang, T. Huang, and J. Luo. Synchronization control for a class of discrete time-delay complex dynamical networks: A dynamic event-triggered approach. *IEEE Trans. Cybern.*, 49(5):1979–1986, 2019.

[95] D. Liberzon. Hybrid feedback stabilization of systems with quantized signals. *Automatica*, 39:1543–1554, 2003.

[96] D. Liberzon. *Switching in Systems and Control.* Birkhauser, Boston, 2003.

[97] D. Liberzon and J.P. Hespanha. Stabilization of nonlinear systems with limited information feedback. *IEEE Trans. Autom. Control*, 50(6):910–915, 2005.

[98] H. Lin and P. Antsaklis. Stability and persistent disturbance attenuation properties for a class of networked control systems: Switched system approach. *Int. J. Control*, 78(18):1447–1458, 2005.

[99] G.-P. Liu, J. X. Mu, D. Rees, and C. Chai. Design and stability analysis of networked control systems with random communication time delay using the modified MPC. *Int. J. Control*, 79(4):288–297, 2006.

[100] G.-P. Liu, Y. Xia, J. Chen, D. Rees, and W. Hu. Design and stability criteria of networked predictive control systems with random network delay in the feedback channel. *IEEE Trans. Syst., Man, Cybern. C*, 37(2):173–184, 2007.

[101] G.-P. Liu, Y. Xia, J. Chen, D. Rees, and W. Hu. Networked predictive control of aystems with random network delays in both forward and feedback channels. *IEEE Trans. Ind. Electron*, 54(3):1282–1297, 2007.

[102] M. Liu, D. W. C. Ho, and Y. Niu. Observer-based controller design for linear systems with limited communication capacity via a descriptor augmentation method. *IET Control Theory Appl.*, 6(3):437–447, 2012.

[103] M. Liu, D. W. C. Ho, and Y. Niu. Observer-based controller design for networked control systems with sensor quantisation and random communication delay. *Int. J. Syst. Sci.*, 43(10):1901–1912, 2012.

[104] T. Liu, Z.-P. Jiang, and D. J. Hill. Decentralized output-feedback control of large-scale nonlinear systems with sensor noise. *Automatica*, 48(10):2560–2568, 2012.

[105] X. Liu and Q. Zhang. New approaches to controller designs based on fuzzy observers for T-S fuzzy systems via LMI. *Automatica*, 39(9):1571–1582, 2003.

[106] J. Lunze and D. Lehmann. A state-feedback approach to event-based control. *Automatica*, 46:211–215, 2010.

[107] S. Luo, F. Deng, and W.-H. Chen. Dynamic event-triggered control for linear stochastic systems with sporadic measurements and communication delays. *Automatica*, 107:86–94, 2019.

[108] M. Mazo Jr., A. Anta, and P. Tabuada. An ISS self-triggered implementation of linear controllers. *Automatica*, 46:1310–1314, 2010.

[109] M. Mazo Jr., A. Anta, and P. Tabuada. Decentralized event-triggered control over wireless sensor/actuator networks. *IEEE Trans. Autom. Control*, 56(10):2456–2461, 2011.

[110] X. Meng, J. Lam, B. Du, and H. Gao. A delay-partitioning approach to the stability analysis of discrete-time systems. *Automatica*, 46(3):610–614, 2010.

[111] Y. Minami, S. Azuma, and T. Sugie. An optimal dynamic quantizer for feedback control with discrete-valued signal constraints. In *Proc. 46th IEEE Conference on Decision and Control*, pages 2259–2264, New Orleans, 2007.

[112] S. K. Nguang. Comments on 'Fuzzy H_∞ tracking control for nonlinear networked control systems in T-S fuzzy model'. *IEEE Trans. Syst., Man, Cybern. B*, 40(3):957–957, 2010.

[113] S. K. Nguang, P. Shi, and S. Ding. Fault detection for uncertain fuzzy systems: an LMI approach. *IEEE Trans. Fuzzy Syst.*, 15(6):1251–1262, 2007.

[114] J. Nilsson, B. Bernhardsson, and B. Wittenmark. Stochastic analysis and control of real-time systems with random time delays. *Automatica*, 34(1):57–64, 1998.

[115] Y. Niu, T. G. Jia, X. Y. Wang, and F. Yang. Output-feedback control design for NCSs subject to quantization and dropout. *Information Sciences*, 179(21):3804–3813, 2009.

[116] P. Otanez, J. Moyne, and D. Tilbury. Using deadbands to reduce communication in networked control systems. In *Proc. American Control Conference*, pages 3015–3020, Anchorage, 2002.

[117] Z.-H. Pang, G.-P. Liu, D. Zhou, and D. Sun. *Networked Predictive Control of Systems with Communication Constraints and Cyber Attacks*. Springer Nature Singapore, Singapore, 2019.

[118] C. Peng and Q.-L. Han. A novel event-triggered transmission scheme and $\mathcal{L}_2$ control co-design for sampled-data control systems. *IEEE Trans. Autom. Control*, 58(10):2620–2626, 2013.

[119] C. Peng, S. Ma, and X. Xie. Observer-based non-PDC control for networked t-s fuzzy systems with an event-triggered communication. *IEEE Trans. Cybern.*, 47(8):2279–2287, 2017.

[120] C. Peng and Y. Tian. Networked H_∞ control of linear systems with state quantization. *Information Sciences*, 177(24):5763–5774, 2007.

[121] C. Peng, Y.-C. Tian, and M. O. Tadé. State feedback controller design of networked control systems with interval time-varying delay and nonlinearity. *Int. J. Robust and Nonlinear Control*, 18(12):1285–1301, 2007.

[122] C. Peng and T. C. Yang. Communication-Delay-Distribution-Dependent networked control for a class of T-S fuzzy systems. *IEEE Trans. Fuzzy Syst.*, 18(2):326–335, 2010.

[123] C. Peng and T. C. Yang. Event-triggered communication and H_∞ control co-design for networked control systems. *Automatica*, 49(5):1326–1332, 2013.

[124] C. Peng, D. Yue, and Q.-L. Han. *Communication and Control for Networked Complex Systems.* Springer-Verlag Berlin, Heidelberg, 2015.

[125] B. Picasso and A. Bicchi. Stabilization of LTI systems with quantized state-quantized input static feedback. In A. Pnueli and O. Maler, editors, *Hybrid Systems: Computation and Control*, volume LNCS 2623 of *Lecture Notes in Computer Science*, pages 405–416. Springer-Verlag, Heidelberg, Germany, 2003.

[126] G. Pin and T. Parisini. Networked predictive control of uncertain constrained nonlinear systems: recursive feasibility and input-to-state stability analysis. *IEEE Trans. Autom. Control*, 56(1):72–87, 2011.

[127] U. Premaratne, S. K. Halgamuge, and I. M. Y. Mareels. Event triggered adaptive differential modulation: A new method for traffic reduction in networked control systems. *IEEE Trans. Autom. Control*, 58(7):1696–1706, 2013.

[128] J. Qiu, G. Feng, and H. Gao. Asynchronous output feedback control of networked nonlinear systems with multiple packet dropouts: T-S fuzzy affine model based approach. *IEEE Trans. Fuzzy Syst.*, 19(6):1014–1030, 2011.

[129] J. Qiu, H. Gao, and S. X. Ding. Recent advances on fuzzy-model-based nonlinear networked control systems: A survey. *IEEE Trans. Ind. Electron.*, 63(2):1207–1217, 2017.

[130] D. R. Quevedo, E. I. Silva, and G. C. Goodwin. Subband coding for networked control systems. *Int. J. Robust Nonlinear Control*, 19:1817–1836, 2009.

[131] K. J. Åström and B. M. Bernhardsson. Comparison of riemann and lebesgue sampling for first order stochastic systems. In *Proc. 41th IEEE Conf. Decision and Control*, pages 2011–2016, Las Vegas, Nevada USA, 2002.

[132] M. Shen, S. Yan, and G. Zhang. A new approach to event-triggered static output feedback control of networked control systems. *ISA Transactions*, 65:468–474, 2016.

[133] Y. Shen, F. Li, D. Zhang, Y.-W. Wang, and Y. Liu. Event-triggered output feedback H_∞ control for networked control systems. *Int. J. Robust Nonlinear Control*, 29:166–179, 2019.

[134] D. Shi, J. Xue, L. Zhao, J. Wang, and Y. Huang. Event-triggered active disturbance rejection control of DC torque motors. *IEEE/ASME Trans. Mechatronics*, 22(5):2277–2287, 2017.

[135] P. Shi. Filtering on sampled-data systems with parametric uncertainty. *IEEE Trans. Autom. Control*, 43(7):1022–1027, 1998.

[136] P. Shi, H. Wang, and C.-C. Lim. Network-based event-triggered control for singular systems with quantizations. *IEEE Trans. Ind. Electron.*, 63(2):1230–1238, 2016.

[137] Y. Shi and B. Yu. Output feedback stabilization of networked control systems with random delays modeled by markov chains. *IEEE Trans. Automat. Control*, 54(7):1668–1674, 2009.

[138] Z. Shu, J. Lam, and J. Xiong. Static output-feedback stabilization of discrete-time markovian jump linear systems: A system augmentation approach. *Automatica*, 46(4):687–694, 2010.

[139] J. Su and W.-H. Chen. Further results on 'Reduced order disturbance observer for discrete-time linear systems'. *Automatica*, 93:550–553, 2018.

[140] J. Sun, J. Yang, S. Li, and W. X. Zheng. Sampled-data-based event-triggered active disturbance rejection control for disturbed systems in networked environment. *IEEE Trans. Cybern.*, 49(2):556–566, 2019.

[141] J. Sun, J. Yang, S. Li, and W. X. Zheng. Output-based dynamic event-triggered mechanisms for disturbance rejection control of networked nonlinear systems. *IEEE Trans. Cybern.*, 50(5):1978–1988, 2020.

[142] X. M. Sun, G.-P. Liu, W. Wang, and R. David. Stability analysis for networked control systems based on average dwell time method. *Int. J. Robust Nonlinear Control*, 20(15):1774–1784, 2010.

[143] P. Tabuada. Event-triggered real-time scheduling of stabilizing control tasks. *IEEE Trans. Autom. Control*, 52(9):1680–1685, 2007.

[144] W. Tan and C. Fu. Linear active disturbance-rejection control: Analysis and tuning via IMC. *IEEE Trans. Ind. Electron.*, 63(4):2350–2359, 2016.

[145] W. Tan and C. Fu. Active disturbance rejection control for active suspension system of tracked vehicles with gun. *IEEE Trans. Ind. Electron.*, 65(5):4051–4060, 2018.

[146] E. Tian, D. Yue, and C. Peng. Quantized output feedback control for networked control systems. *Information Sciences*, 178(12):2734–2749, 2008.

[147] E. Tian, D. Yue, and X. Zhao. Quantised control design for networked control systems. *IET Control Theory Appl.*, 1(6):1693–1699, 2007.

[148] Y. Tipsuwan and M. Y. Chow. Control methodologies in networked control systems. *Control Eng. Practice*, 11:1099–1111, 2003.

[149] H. Trinh and T. Fernando. *Functional observers for dynamical systems*. Springer, Berlin, 2012.

[150] N. S. Tripathy, I. N. Kar, and K. Paul. Robust dynamic event-triggered control for linear uncertain system. *IFAC-Papers OnLine*, 49(1):207–212, 2016.

[151] C.-C. Ku, W.-J. Chang, and P.-H. Huang. Robust fuzzy control via observer feedback for passive stochastic fuzzy systems with time-delay and multiplicative noise. *Int. J. of Innovative Computing, Information and Control*, 7(1):345–364, 2011.

[152] D. Wang, , J. Wang, and W. Wang. Output feedback control of networked control systems with packet dropouts in both channel. *Information Sciences*, 221:544–554, 2013.

[153] F.-Y. Wang and D. Liu. *Networked Control Systems: Theory and Applications*. Springer-Verlag, London, 2008.

[154] R. Wang, G.-P. Liu, B. Wang, W. Wang, and D. Rees. l_2-gain analysis for networked predictive control systems based on switching method. *Int. J. Control*, 82(6):1148–1156, 2009.

[155] R. Wang, G.-P. Liu, W. Wang, D. Rees, and Y.-B. Zhao. Guaranteed cost control for networked control systems based on an improved predictive control method. *IEEE Trans. Control Syst. Technol.*, 18(5):1226–1232, 2010.

[156] R. Wang, G.-P. Liu, W. Wang, D. Rees, and Y.-B. Zhao. H_∞ control for networked predictive control systems based on the switched lyapunov function method. *IEEE Trans. Ind. Electron.*, 57(10):3565–3571, 2010.

[157] X. Wang and M. Lemmon. Self-triggered feedback control systems with finite-gain $\mathcal{L}_2$ stability. *IEEE Trans. Autom. Control*, 54(3):452–467, 2009.

[158] X. Wang and M. Lemmon. Self-triggering under state-independent disturbances. *IEEE Trans. Autom. Control*, 55(6):1494–1500, 2010.

[159] Y. Wang, Z. Jia, and Z. Zuo. Dynamic event-triggered and self-triggered output feedback control of networked switched linear systems. *Neurocomputing*, 314:39–47, 2018.

[160] Y. Wang, W. X. Zheng, and H. Zhang. Dynamic event-based control of nonlinear stochastic systems. *IEEE Trans. Autom. Control*, 62(12):6544–6551, 2017.

[161] Y.-L. Wang, C.-C. Lim, and P. Shi. Adaptively adjusted event-triggering mechanism on fault detection for networked control systems. *IEEE Trans. Cybern.*, 47(8):2299–2311, 2017.

[162] Y.L. Wang and G.-H. Yang. Output tracking control for networked control systems with time delay and packet dropout. *Int. J. Control*, 81(11):1709–1719, 2008.

[163] Z. Wang, F. Yang, D. W. C. Ho, and X. Liu. Robust H_∞ control for networked systems with random packet losses. *IEEE Trans. Syst., Man, Cybern. B, Cybern.*, 37(4):916–924, 2007.

[164] R. L. Williams and D. A. Lawrence. *Linear State-Space Control Systems*. John Wiley & Sons, Inc., 2007.

[165] J. Wu and T. Chen. Design of networked control systems with packet dropouts. *IEEE Trans. Autom. Control*, 52(7):1314–1319, 2007.

[166] J. Wu, H. R. Karimi, and P. Shi. Network-based H_∞ output feedback control for uncertain stochastic systems. *Information Sciences*, 2013. In Press.

[167] L. Wu, J. Lam, H. Gao, and J. Xiong. Robust guaranteed cost control of discrete-time networked control systems. *Optimal Control Applications and Methods*, 32(1):95–112, 2011.

[168] L. Wu, X. Su, P. Shi, and J. Qiu. Model approximation for discrete-time state-delay systems in the T-S fuzzy framework. *IEEE Trans. Fuzzy Syst.*, 19(2):366–378, 2011.

[169] L. Wu, X. Su, P. Shi, and J. Qiu. A new approach to stability analysis and stabilization of discrete-time T-S fuzzy time-varying delay systems. *IEEE Trans. Syst., Man, Cybern. B*, 41(1):273–286, 2011.

[170] Z.-G. Wu, Y. Xu, Y.-J. Pan, P. Shi, and Q. Wang. Event-triggered pinning control for consensus of multiagent systems with quantized information. *IEEE Trans. Syst., Man, Cybern., Syst.*, 48(11):1929–1938, 2018.

[171] Y. Xia, M. Fu, and G.-P. Liu. *Analysis and Synthesis of Networked Control Systems*. Springer-Verlag, Berlin Heidelberg, 2011.

[172] Y. Xia, G.-P. Liu, P. Shi, D. Rees, and E. J. C. Thomas. New stability and stabilization conditions for systems with time-delay. *Int. J. Syst. Sci.*, 38(1):17–24, 2007.

[173] Y. Xia, J. Zhang, and E. K. Boukas. Control for discrete singular hybrid systems. *Automatica*, 44(10):2635–2641, 2008.

[174] L. Xiao, A. Hassibi, and J. P. How. Control with random communication delays via a discrete-time jump system approach. In *Proc. Amer. Control Conf.*, Chicago, IL, 2000.

[175] L. Xie and L. H. Xie. Stability analysis and stabilization of networked linear systems with random packet losses. *Sci. China Ser. F: Inf. Sci.*, 52(11):2053–2073, 2009.

[176] J. Xiong and J. Lam. Stabilization of linear systems over networks with bounded packet loss. *Automatica*, 43(1):80–87, 2007.

[177] J. Xiong and J. Lam. Stabilization of networked control systems with a logic ZOH. *IEEE Trans. Autom. Control*, 54(2):358–363, 2009.

[178] S. Xu and J. Lam. *Robust Control and Filtering of Singular Systems.* Springer-Verlag, Berlin, 2006.

[179] H. Yang, X. Guo, L. Dai, and Y. Xia. Event-triggered predictive control for networked control systems with network-induced delays and packet dropouts. *Int. J. Robust and Nonlinear Control,* 28:1350–1365, 2018.

[180] H. Yang, Y. Xia, and P. Shi. Stabilization of networked control systems with nonuniform random sampling periods. *Int. J. Robust and Nonlinear Control,* 21(5):501–526, 2011.

[181] R. Yang, G.-P. Liu, P. Shi, C. Thomas, and M. V. Basin. Predictive output feedback control for networked control systems. *IEEE Trans. Ind. Electron.,* 61(1):512–520, 2014.

[182] R. Yang, P. Shi, and G.-P. Liu. H_∞ filtering for discrete-time networked nonlinear systems with mixed random delays and packet dropouts. *IEEE Trans. Autom. Control,* 56(11):2655–2660, 2011.

[183] X. Yi, K. Liu, D. V. Dimarogonas, and K. H. Johansson. Distributed dynamic event-triggered control for multi-agent systems. In *Proc. of IEEE 56th Annual Conference on Decision and Control,* pages 6683–6688, Melbourne, Australia, 2017.

[184] K. You, N. Xiao, and L. Xie. *Analysis and Design of Networked Control Systems.* Springer-Verlag London, London, 2015.

[185] H. Yu and P. J. Antsaklis. Event-triggered output feedback control for networked control systems using passivity: Achieving L_2 stability in the presence of communication delays and signal quantization. *Automatica,* 49(1):30–38, 2013.

[186] D. Yue, Q.-L. Han, and J. Lam. Network-based robust H_∞ control of systems with uncertainty. *Automatica,* 41(6):999–1007, 2005.

[187] D. Yue, Q.-L. Han, and C. Peng. State feedback controller design of networked control systems. *IEEE Trans. Circuits Syst. II, Exp. Briefs,* 51(11):640–644, 2004.

[188] D. Yue, C. Peng, and G. Yang. Guaranteed cost control of linear systems over networks with state and input quantisations. *IEE Proc. Control Theory and Applications,* 153:658–664, 2006.

[189] D. Yue, E. Tian, and Q.-L. Han. A delay system method for designing event-triggered controllers of networked control systems. *IEEE Trans. Autom. Control,* 58(2):475–481, 2013.

[190] D. Yue, E. Tian, Z. Wang, and J. Lam. Stabilization of systems with probabilistic interval input delays and its applications to networked control systems. *IEEE Trans. Syst., Man Cybern. A, Syst., Humans,* 39(4):939–945, 2009.

[191] D. Yue, E. Tian, Y. Zhang, and C. Peng. Delay-distribution-dependent stability and stabilization of T-S fuzzy systems with probabilistic interval delay. *IEEE Trans. Syst., Man Cybern. B, Cybern.*, 39(2):503–516, 2009.

[192] C. Zhan, Y. Fa, C. Son, Y. Yan, and M. Shen. Event-triggered control with output feedback using dynamic triggering mechanism. In *Proc. of the 32nd Youth Academic Annual Conference of Chinese Association of Automation*, pages 346–351, Hefei, China, 2017.

[193] D. Zhang, Q.-L. Han, and X.-C. Jia. Network-based output tracking control for T-S fuzzy systems using an event-triggered communication scheme. *Fuzzy Sets Syst.*, 273:26–48, 2015.

[194] H. Zhang, G. Feng, H. Yan, and Q. Chen. Observer-based output feedback event-triggered control for consensus of multi-agent systems. *IEEE Trans. Ind. Electron.*, 61(9):4885–4894, 2014.

[195] H. Zhang, Y. Shi, and M. Liu. H_∞ step tracking control for networked discrete-time nonlinear systems with integral and predictive actions. *IEEE Trans. Ind. Inf.*, 9(1):337–345, 2013.

[196] H. Zhang, Y. Shi, and A. S. Mehr. Robust weighted H_∞ filtering for networked systems with intermittent measurements of multiple sensors. *International Journal of Adaptive Control and Signal Processing*, 25(4):313–330, 2010.

[197] H. Zhang, J. Yang, and C.-Y. Su. T-S fuzzy-model-based robust H_∞ design for networked control systems with uncertainties. *IEEE Trans. Ind. Informat.*, 3(4):289–301, 2007.

[198] J. Zhang and G. Feng. Event-driven observer-based output feedback control for linear systems. *Automatica*, 50(7):1852–1859, 2014.

[199] J. Zhang, J. Lam, and Y. Xia. Observer-based output feedback control for discrete systems with quantized inputs. *IET Control Theory & Applications*, 5(3):478–485, 2011.

[200] J. Zhang, J. Lam, and Y. Xia. Output feedback sliding mode control under networked environment. *Int. J. Syst. Sci.*, 44(4):750–759, 2013.

[201] J. Zhang, J. Lam, and Y. Xia. Output feedback networked predictive control for systems with random network delays. *Information Sciences*, 265:154–166, 2014.

[202] J. Zhang, Y. Lin, and P. Shi. Output tracking control of networked control systems via delay compensation controllers. *Automatica*, 57:85–92, 2015.

[203] J. Zhang, P. Shi, and Y. Xia. Robust adaptive sliding mode control for fuzzy systems with mismatched uncertainties. *IEEE Trans. Fuzzy Sys.*, 18(4):700–711, 2010.

[204] J. Zhang, P. Shi, and Y. Xia. Fuzzy delay compensation control for T-S fuzzy systems over network. *IEEE Trans. Cybern.*, 43(1):259–268, 2013.

[205] J. Zhang, P. Shi, Y. Xia, and H. Yang. Discrete-time sliding mode control with disturbance rejection. *IEEE Trans. Ind. Electron.*, 66(10):7967–7975, 2019.

[206] J. Zhang and Y. Xia. Design of H_∞ fuzzy controllers for nonlinear systems with random data dropouts. *Optimal Control, Applications and Methods*, 32(3):328–349, 2011.

[207] J. Zhang, Y. Xia, and P. Shi. Design and stability analysis of networked predictive control systems. *IEEE Trans. Control Syst. Technol.*, 21(4):1495–1501, 2013.

[208] J. Zhang, W. X. Zheng, H. Xu, and Y. Xia. Observer-based event-driven control for discrete-time systems with disturbance rejcetion. *IEEE Trans. Cybern.*, 51(4):2120–2130, 2021.

[209] L. Zhang, Y. Shi, T. Chen, and B. Huang. A new method for stabilization of networked control systems with random delays. *IEEE Trans. Autom. Control*, 50(8):1177–1181, 2005.

[210] L. Zhang, C. Wang, and L. Chen. Stability and stabilization of a class of multimode linear discrete-time systems with polytopic uncertainties. *IEEE Trans. Ind. Electron.*, 56(9):3684–3692, 2009.

[211] W. Zhang. *Stability analysis of networked control systems*. PhD thesis, Case Western Reserve University, Department of Electrical Engineering and Computer Science, 2001.

[212] W.-A. Zhang and L. Yu. Output feedback stabilization of networked control systems with packet dropouts. *IEEE Trans. Autom. Control*, 52(9):1705–1710, 2007.

[213] W.-A. Zhang and L. Yu. Modelling and control of networked control systems with both network-induced delay and packet-dropout. *Automatica*, 44(12):3206–3210, 2008.

[214] W.-A. Zhang and L. Yu. A robust control approach to stabilization of networked control systems with time-varying delays. *Automatica*, 45(10):2440–2445, 2008.

[215] W.-A. Zhang and L. Yu. Stability analysis for discrete-time switched time-delay systems. *Automatica*, 45(10):2265–2271, 2009.

[216] W.-A. Zhang and L. Yu. Stabilization of sampled-data control systems with control inputs missing. *IEEE Trans. Autom. Control*, 55(2):447–452, 2010.

[217] W.-A. Zhang, L. Yu, and H. Song. H_∞ filtering of networked discrete-time systems with random packet losses. *Information Sciences*, 179(22):3944–3955, 2009.

[218] X. Zhang, G. Lu, and Y. Zheng. Stabilization of networked stochastic time-delay fuzzy systems with data dropout. *IEEE Trans. Fuzzy Syst.*, 16(3):798–807, 2008.

[219] X.-M. Zhang and Q.-L. Han. Event-triggered dynamic output feedback control for networked control systems. *IET Control Theory Appl.*, 8(4):226–234, 2014.

[220] X.-M. Zhang, Q.-L. Han, and B.-L. Zhang. An overview and deep investigation on sampled-data-based event-triggered control and filtering for networked systems. *IEEE Trans. Ind. Inform.*, 13(1):4–16, 2017.

[221] X.-M. Zhang, M. Wu, J.-H. She, and Y. He. Delay-dependent stabilization of linear systems with time-varying state and input delays. *Automatica*, 41(8):1405–1412, 2005.

[222] Z. Zhang, L. Zhang, F. Hao, and L. Wang. Periodic event-triggered consensus with quantization. *IEEE Trans. Circuits Syst. II, Exp. Briefs*, 63(4):406–410, 2016.

[223] Y.-B. Zhao, J. Kim, and G.-P. Liu. Error bounded sensing for packet-based networked control systems. *IEEE Trans. Ind. Electron.*, 58(5):1980–1989, 2011.

[224] Y.-B. Zhao, G.-P. Liu, Y. Kang, and L. Yu. *Packet-Based Control for Networked Control Systems*. Science Press, Beijing and Springer Nature Singapore, Singapore, 2018.

[225] Y.-B. Zhao, G.-P. Liu, and D. Rees. Networked predictive control systems based on hammerstein model. *IEEE Trans. Circuits Syst. II-Express Briefs*, 55(5):469–473, 2008.

[226] Y.-B. Zhao, G.-P. Liu, and D. Rees. A predictive control-based approach to networked hammerstein systems: Design and stability analysis. *IEEE Trans. Syst., Man, Cybern. B, Cybern.*, 38(3):700–708, 2008.

[227] Y.-B. Zhao, G.-P. Liu, and D. Rees. A predictive control based approach to networked wiener systems. *Int. J. Innov. Comp. Inf. Control*, 4(11):2793–2802, 2008.

[228] Y.-B. Zhao, G.-P. Liu, and D. Rees. Design of a packet-based control framework for networked control systems. *IEEE Trans. Control Syst. Technol.*, 17(4):859–865, 2009.

[229] Y.-B. Zhao, G.-P. Liu, and D. Rees. Packet-based deadband control for internet-based networked control systems. *IEEE Trans. Control Syst. Technol.*, 18(5):1057–1067, 2010.

[230] Q. Zheng, L. Dong, D. H. Lee, and Z. Gao. Active disturbance rejection control for MEMS gyroscopes. *IEEE Trans. Control Syst. Technol.*, 17(6):1432–1438, 2009.

[231] Q. Zhou and P. Shi. A new approach to network-based H_∞ control for stochastic systems. *Int. J. Robust and Nonlinear Control*, 22(9):1036–1059, 2012.

[232] Y. Zou, Y. Niu, B. Chen, and T. Jia. Networked predictive control of constrained linear systems with input quantisation. *Int. J. Syst. Science*, 44(10):1970–1982, 2013.

[233] Z. Zuo, S. Guan, Y. Wang, and H. Li. Dynamic event-triggered and self-triggered control for saturated systems with anti-windup compensation. *J. Frank. Inst.*, 354(17):7624–7642, 2017.

Index

active disturbance rejection control (ADRC), 153
asymptotically stable, 148
average dwell time, 38, 44

BIBO stable, 90
bilinear matrix inequalities (BMIs), 31

CCL algorithm, 46
comparison lemma, 100, 127
cone complementarity linearization (CCL), 18
continuous event-triggered control, 6
control input predictions (CIPs), 37
controllable, 14, 98
controller-to-actuator channel delay, 39
cyber-physical systems (CPSs), 1

data dropout, 1, 13
decoder, 139
detectable, 85, 112, 157
dynamic event-triggered control, 7
dynamic output feedback (DOF), 125

encoder, 139
event threshold, 99
event-triggered ADRC, 153
event-triggered communication, 2
event-triggered control, 1, 97
event-triggered quantizer, 140
exponential stability, 46
exponentially stable, 42
extended functional state observer (EFSO), 81, 153
extended state observer (ESO), 153

ILMI algorithm, 46
independent and identically-distributed, 14
inter-event interval, 126

linear matrix inequalities (LMIs), 31
linear quadratic regulator (LQR), 40

Markov chain, 54
Markovian jump system, 52
membership degree, 69
minimum inter-event interval, 99, 101, 128

network communication delay, 1, 13
networked control, 1
networked delay compensation control, 37

observable, 14, 98

parallel distributed compensation (PDC), 67
PD controller, 15
periodic event-triggered control, 6
pole assignment, 40, 92, 107
predictive vector quantizer (PVQ), 139

quadratic stabilizability, 25
quantization density, 27
quantized event-triggered controller, 139
quantizer, 26

reduced-order extended state functional observer, 166
Riccati equation, 144, 150
round trip delay, 2, 14, 38, 40, 53, 69, 70

Schur complement lemma, 18, 30, 44, 55, 130
Schur stable, 40
self-triggered control, 7, 111
sensor-to-controller channel delay, 39
signal quantization, 1, 25
singular fuzzy system, 73
singular Markovian jump system, 52

stabilizable, 112
stabilization, 13
static output feedback control, 125
steady-state, 160
stochastic stability, 13
stochastic system, 4
stochastically admissible, 58
stochastically stable, 16, 55
switched system, 3, 38
switched time delay system, 41
switching signal, 41

Takagi-Sugeno (T-S) fuzzy model, 67
time delay system, 2, 41
time stamp, 14
transition probability matrix, 54

Wirtinger inequality, 132

Zeno behavior, 6
zero-order hold (ZOH), 99
Z-transform, 160